Advanced Environmental Monitoring with Remote Sensing Time Series Data and R

T0331549

Advanced Environmental Monitoring with Remote Sensing Time Series Data and R

Alexandra Gemitzi, Nikolaos Koutsias, and Venkat Lakshmi

CRC Press

Taylor & Francis Group

Boca Raton London New York

CRC Press is an imprint of the
Taylor & Francis Group, an **informa** business

CRC Press
Taylor & Francis Group
6000 Broken Sound Parkway NW, Suite 300
Boca Raton, FL 33487–2742

© 2020 by Taylor & Francis Group, LLC

CRC Press is an imprint of Taylor & Francis Group, an Informa business

**Visit the Taylor & Francis Web site at
www.taylorandfrancis.com**

**and the CRC Press Web site at
www.crcpress.com**

Contents

Preface

EARTH IS A REMARKABLE planet. Not only because it supports life but also because of its ability to react to external and internal stresses. Earthquakes, volcanic eruptions, storm surges, landslides, wildfires, and tides are among the numerous natural phenomena that affect Earth and cause changes at various scales. Throughout Earth's history tremendous changes have happened and have altered the environment and climate of the planet. From the Carboniferous period with the boosting vegetation to the Permian-Triassic extinction event and from the Late Triassic with the arrival of dinosaurs to the Cretaceous-Paleogene extinction event and to the Cenozoic mammal evolution, Earth experienced unbelievable changes that brought dramatic impacts on life. Climate changes, with warming and cooling periods, have affected Earth many times in the past and are nowadays accelerating due to human activities. No one can decide on the positive or negative character of those changes through geological history. But the one thing all of us can be sure of is that Earth will continue to change, as any other living creature, until the unavoidable end. Life is tightly connected to the environment, and any changes on it are reflected on living organisms. From the beginning of human history, people explored Earth to acquire knowledge related to the evolution of life and the environment. Despite the efforts devoted to environmental monitoring, only recently did humans manage to achieve a complete picture of Earth and its environment, as global scale observations were restricted by technological constraints. Recent advances of space, satellite, and information technology facilitated global observations that can detect even minor changes not only on Earth's atmosphere and surface but also in the subsurface. Results indicate, beyond any doubt, the changing nature of our Earth. It remains on science to answer the question of how those changes can contribute to the benefit of life and on policy makers to adopt sustainable solutions for the future of our planet.

Authors

Alexandra Gemitzi graduated from Aristotle University of Thessaloniki (Greece) in 1993 with a Bachelor's degree in Geology and a Master of Science in Groundwater Engineering from University of Newcastle upon Tyne (UK) in 1995. She obtained a Doctorate from the Department of Civil Engineering of Aristotle University of Thessaloniki, sponsored by the National Scholarship Foundation. She is Associate Professor at the Department of Environmental Engineering of Democritus University of Thrace, Greece. Before starting her academic career, she worked as a geologist for almost ten years in the private and public sector. Her main teaching and research interests are related to environmental modeling, remote sensing, and Geographic Information Systems. Her research focuses on the simulation and assessment of environmental systems, climate and land use change effects, and the prediction and risk assessment of extreme natural phenomena or natural disasters. Her published work features over 30 publications in peer-reviewed journals and over 50 conference presentations.

Nikolaos (Nikos) Koutsias is Associate Professor at the Department of Environmental Engineering of University of Patras, where he gives courses in informatics, remote sensing, Geographic Information Systems (GIS), and spatial analysis. He obtained his diploma degree in Environmental Studies from the University of Aegean, his M.Sc. degree in Environmental and Renewable Resources from the Mediterranean Agronomic Institute of Chania (MAICh), and his Ph.D. in Remote Sensing of Burned Land Mapping from the Department of Forestry and Natural Environment, Aristotle University of Thessaloniki. He had a post-doctorate position at the GIS division in the Department of Geography of the University of Zurich. He has been working in the field of remote sensing and GIS with special emphasis on wildland fires, forestry, and ecology for the last 25 years. He has participated in national and European projects and has several publications in journals, books, and international conferences. He is a recent recipient of a Marie Curie Individual Scholarship.

Venkat Lakshmi graduated from University of Roorkee in 1987 with a Bachelor's degree in Civil Engineering and from Princeton in 1997 with a Doctorate in Civil and Environmental Engineering. He worked at NASA Goddard Space Flight Center 1996–1999 as a research scientist in the Laboratory for the Atmospheres. His areas of research interest are catchment hydrology, satellite data validation and assimilation, field experiments, land-atmosphere interactions, satellite data downscaling, vadose zone, and water resources. He has served as the Carolina Trustee Professor and as a former chair of the Department of Earth and Ocean Sciences at the University of South Carolina (1999–2018) and is currently Professor in the Department of Engineering Systems and the Environment at the University of Virginia. He has served as Cox Visiting Professor at Stanford University 2006–2007 and 2015–2016 and Program Director for Hydrologic Sciences at the National Science Foundation (2017–2018). Venkat is a fellow of the American Society of Civil Engineers (ASCE), and he has over 100 peer-reviewed articles and 300 presentations and has served as thesis supervisor for 25 graduate students. He is currently is serving as Associate Editor of *Journal of Hydrology* and as an editor for *Vadose Zone Journal* and is a former chairman of the Chapman Conference committee for American Geophysical Union (AGU) and the founding editor-in-chief of *Remote Sensing in Earth System Science* (Springer Journals). He has served on the National Academies Panel for the Decadal Survey of Earth Observations from Space (NASA) and as chair of the planning committee for Groundwater Recharge and Flow: Approaches and Challenges for Monitoring and Modeling Using Remotely Sensed Data (NGA).

ADDITIONAL CONTRIBUTIONS FROM

Maria A. Banti
John Bolten
Bin Fang
Ibrahim N. Mohammed

Introduction

Alexandra Gemitzi

R EMOTE SENSING, IN ITS broader meaning, is the process of defining the physical characteristics of an area by means of a flying device, either aircraft or satellite. Most of the time we consider remote sensing as a technique that measures the reflected and emitted radiation of Earth's surface. However, remote sensing also consists of methods that are not related to monitoring of any kind of radiation, but they focus on indirectly detecting some physical properties of Earth such as gravitational and magnetic field. Those latter have been mostly part of geophysical surveying, traditionally being performed on the ground using hand-held instruments. Nowadays remote sensing measurements of gravitational and magnetic field have become very popular and are considered as standard remotely sensed techniques as well.

Remotely sensed systems can be categorized based on the detection properties of sensors (sensitivity to various spectral areas like optical, infrared, or microwave radiation), the way they monitor radiation (passive or active), the orbit altitude, and the type of orbit (e.g. geostationary or polar). Other important factors are various types of resolution, e.g. spatial resolution, temporal resolution, radiometric resolution, spectral resolution. Each system provides products for specific applications, e.g. most of the meteorological satellites operate at an altitude of 800 to 1,500 km, on a polar or near polar sun synchronous orbit. Geostationary orbiting satellites with a much higher altitude of 36,000 km rotate at a speed of approximately 3 km/sec, same as the rotational speed of Earth, and therefore they seem to be stationary. Their applications are mostly focused on telecommunications, but they are also very useful at monitoring continent wide weather and environmental conditions.

Our book focuses on satellite missions devoted to global environmental monitoring. Those are of polar or near polar sun synchronous orbits, and they usually operate at an altitude of 600 to 800 km, providing improved ground observations, as the target area is viewed each revisited time at the same illumination conditions especially concerning the angle of the incoming Sun radiation. Nevertheless, the wide applicability of each remote sensing product is dependent not only on merely scientific factors, like those mentioned prior, but also on the cost and the ease of accessibility of the available data sets. We have to admit that we are lucky enough, not only as scientists but also as individuals, to have access to a vast quantity of free remotely sensed environmental information. What one really needs to know is where to find it through the numerous available web services and perhaps most importantly how to process all this information. Our book will guide the reader through the procedures of accessing and handling the freely available state of the art remotely sensed products and take advantage of the open data access platforms. Our work focuses on global environmental monitoring missions that have been operating for at least two decades and that have undergone rigorous reliability checks, which nowadays provide a long enough time series suitable for unveiling global but also regional and local environmental changes. Categorization of those products was based on their spatial and temporal resolution as those factors play a crucial role on their applicability to various environmental assessments.

## CHAPTER 1

Environmental Application of Medium to High Resolution Remotely Sensed Data

Nikolaos Koutsias

1.1 INTRODUCTION: BACKGROUND ON THE SPECTRAL INFORMATION OF MEDIUM TO HIGH RESOLUTION SENSORS

Landsat Series

The Landsat series has a long history of data capture starting with the launch of Landsat 1 on July 23, 1972, with the Multispectral Scanner (MSS) onboard (originally known as Earth Resources Technology Satellite [ERTS]). Since that time, Landsat satellites have been taking repetitive images of Earth's surface at continental scale, thus creating a huge historical archive that can be used to reconstruct the past (Nioti, Dimopoulos, and Koutsias 2011). Landsat 1, 2, and 3 carried the MSS, whereas Landsat 4 and 5 carried the onboard Thematic Mapper (TM), an advanced multispectral scanning radiometer that had higher spatial and spectral resolution than Landsat 1–3 and an improved geometric accuracy. Following the failure of Landsat 6, Landsat 7 was successfully launched on April 15, 1999, carrying the Enhanced Thematic Mapper Plus (ETM+). ETM+ data offer seven bands as TM which cover the visible, near, short-wave, and thermal infrared part of the electromagnetic spectrum and additionally a high resolution panchromatic band (for the first time in the Landsat series). The spatial resolution ranges from 15 to 60 meters.

Landsat 8 (OLI, Operational Land Imager, and TIRS, Thermal Infrared Sensor) was launched on February 11, 2013, and has three new bands along with two additional thermal bands. Specifically, two spectral bands correspond to (i) a deep blue visible channel (band 1) specifically designed for water resources and coastal zone investigation, and (ii) a new infrared channel (band 9) for the detection of cirrus clouds. Additionally, a new Quality Assurance band is also included with each data product providing information on the presence of features such as clouds, water and snow. Finally, the TIRS instrument collects two spectral bands in the thermal infrared region of the wavelength as compared to a single band on the previous TM and ETM+ sensors (http://landsat.usgs.gov/). Finally, the new Landsat-8 OLI sensor has a higher radiometric resolution of 16-bit as compared to the 8-bit resolution provided by Landsat-7. This enhanced feature of Landsat-8 may enable greater sensitivity and reliability in the mapping and monitoring of burned areas. The Landsat spectral band specifications, including the upcoming Landsat-9, are listed in Table 1.1, while graphically are presented in Figure 1.1.

TABLE 1.1 The Landsat spectral bands specifications

Band name	Band number	Wavelength		Resolution
		Landsat Multispectral Scanner (MSS)		
	Landsat MSS 1, 2, 3 /Landsat MSS 4, 5	Landsat 1, 2, 3 MSS	Landsat 4, 5 MSS	
Green	˙1–3 (4)/4–5 (1)	0.5–0.6	0.5–0.6	68 × 83 m
Red	1–3 (5)/4–5 (2)	0.6–0.7	0.6–0.7	68 × 83 m
Near Infrared (NIR)	1–3 (6)/4–5 (3)	0.7–0.8	0.7–0.8	68 × 83 m
NIR	1–3 (7)/4–5 (4)	0.8–1.1	0.8–1.1	68 × 83 m
Thermal	8 (Landsat-3)	10.4–12.6 (Landsat-3)		68 × 83 m
		Landsat 4–5 Thematic Mapper (TM) and Landsat 7 Enhanced Thematic Mapper Plus (ETM+)		
		Landsat 4/5 TM	Landsat-7 ETM+	
Blue	1	0.45–0.52	0.45–0.52	30 m
Green	2	0.52–0.60	0.52–0.60	30 m
Red	3	0.63–0.69	0.63–0.69	30 m
NIR	4	0.76–0.90	0.77–0.90	30 m
Shortwave Infrared (SWIR) 1	5	1.55–1.75	1.55–1.75	30 m
Thermal	6	10.41–12.5	10.40–12.50	120 m
SWIR 2	7	2.08–2.35	2.09–2.35	30 m
Panchromatic	8		0.52–0.90	
		Landsat 8/9 Operational Land Image (OLI/OLI-2) and Thermal Infrared Sensor (TIRS/TIRS-2)		
		Landsat-8 OLI/TIRS	Landsat-9 OLI-2/TIRS-2	
Coastal aerosol	1	0.433–0.453	0.433–0.453	30 m
Blue	2	0.450–0.515	0.45–0.515	30 m
Green	3	0.525–0.600	0.525–0.6	30 m
Red	4	0.630–0.680	0.63–0.68	30 m
NIR	5	0.845–0.885	0.845–0.885	30 m
SWIR 1	6	1.560–1.660	1.56–1.66	30 m
SWIR 2	7	2.100–2.300	2.1–2.3	30 m
Panchromatic	8	0.500–0.680	0.5–0.68	15 m
Cirrus	9	1.360–1.390	1.36–1.39	30 m
Thermal	10	10.6–11.2	10.3–11.3	100 m
Thermal	11	11.5–12.5	11.5–12.5	100 m

Sentinel-2 Series

European Space Agency (ESA) is currently developing a new family of missions called Sentinels, which includes seven missions providing radar and super-spectral imaging for land, ocean, and atmospheric monitoring. Each Sentinel mission is based on a constellation of two satellites to revisit the coverage requirements and provide robust data sets for Copernicus Services. Among them Sentinel-2 is a polar-orbiting, multispectral high resolution imaging mission that comprises a constellation of two twin polar-orbiting satellites (Sentinel-2A and Sentinel-2B) placed in the same sun synchronous orbit, phased at 180 degrees to each other. Sentinel-2A was launched on June 23, 2015, and Sentinel-2B followed on March 7, 2017. The two sensors provide satellite data every five days at the equator under cloud-free conditions to monitor vegetation, soil and water cover, inland waterways, and coastal areas.

Sentinel-2 carries an optical instrument payload that samples 13 spectral bands at 10 m (four bands), at 20 m (six bands), and at 60 m (three bands) spatial resolution with an orbital swath width of 290 km. The twin satellites of Sentinel-2 continue data collection of SPOT (Satellite Pour l'Observation de la Terre) and LANDSAT-type image data, contributing thus to ongoing multispectral observations and used for various applications including among others land management, agriculture and forestry, disaster control, humanitarian relief operations, risk mapping, and security concerns.

Sentinel-2, among other improved features over the more classical Landsat and SPOT satellite sensors, has a few red-edge bands (5, 6, and 7, although band 7 overlaps with NIR band 8) and a narrow NIR band. This unique feature of the Sentinel-2 sensor provides more capabilities especially for studies concerning vegetation and therefore also monitoring burned surfaces. Actually, red-edge spectral indices based on B5 and B7/B8 showed the highest suitability for burn severity discrimination (Fernández-Manso, Fernández-Manso, and Quintano 2016). Additionally, based on Huang et al. (2016), the most suitable Multi-Spectral Instrument (MSI) bands to detect burned areas are the 20 m near-infrared,

TABLE 1.2 The Sentinel-2 spectral bands specifications

Band name	Band number	Sentinel-2A		Sentinel-2B		Spatial resolution (m)
		Central wavelength (nm)	Bandwidth (nm)	Central wavelength (nm)	Bandwidth (nm)	
Coastal aerosol	1	442.7	21	442.2	21	60
Blue	2	492.4	66	492.1	66	10
Green	3	559.8	36	559.0	36	10
Red	4	664.6	31	664.9	31	10
Vegetation red edge	5	704.1	15	703.8	16	20
Vegetation red edge	6	740.5	15	739.1	15	20
Vegetation red edge	7	782.8	20	779.7	20	20
NIR	8	832.8	106	832.9	106	10
Narrow NIR	8A	864.7	21	864.0	22	20
Water vapour	9	945.1	20	943.2	21	60
SWIR – Cirrus	10	1373.5	31	1376.9	30	60
SWIR	11	1613.7	91	1610.4	94	20
SWIR	12	2202.4	175	2185.7	185	20

shortwave infrared, and red-edge bands, while the performance of the spectral indices varied with location. The Sentinel-2 spectral band specifications are listed in Table 1.2, while all original spectral channels and RGB composites that arise from the correspondence of red to SWIR channel, green to NIR channels, and blue to visible channels, that enhance the discrimination of burned surfaces, are provided in Figure 1.2.

A diagram showing comparatively the spectral bands of Landsat and Sentinel-2 satellite sensors (excluding the thermal bands) along with some spectral curves captured by a portable field spectroradiometer (ASD FieldSpec® 4 Hi-Res) of vegetated, dry-vegetated, and burned areas is presented in Figure 1.1.

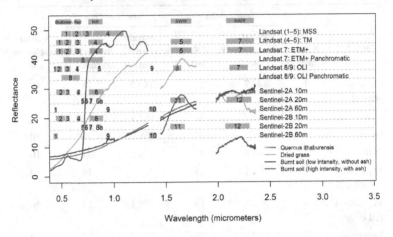

FIGURE 1.1 Position of the spectral bands (excluding the thermal) of Landsat and Sentinel-2 satellite sensors in the electromagnetic spectrum along with some spectral curves captured by a portable field spectroradiometer (ASD FieldSpec® 4 Hi-Res) of vegetated, dry-vegetated, and burned areas.

Spectral Properties of Burned Surfaces: The Basis of Their Assessment and Monitoring

The spectral information of the objects in remote sensing is depicted in the spectral signatures or spectral response curves which is the interaction between the objects and the electromagnetic radiation. The spectral information of the objects is the basis of the remote sensing, and depending on how well the objects are spectrally separated they can be successfully mapped and monitored with satellite remote sensing data.

The spectral properties of burned surfaces have been studied and explored very well in Landsat satellite images, and their spectral patterns are very well known from many studies (Pereira et al. 1997). Two main spectral patterns have been reported; one that is defined by the spectral behavior in the visible channels and one that is defined by the infrared channels. These spectral

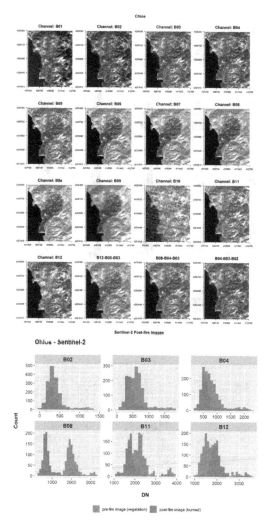

FIGURE 1.2   Spectral channels of Sentinel-2 sensor along with few RGB color composites (left upper panel), RGB color composites from SWIR-NIR-Visible channels (right upper panel), histogram data plots of burned (post-fire image) vs. non-burned samples (pre-fire image) (left lower panel), and spectral signatures plots of various land cover types in the pre- and post-fire Sentinel-2 image (right lower panel). The two lower graphs concern only the B02, B03, B04, B08, B11, and B12 channels.

(continued)

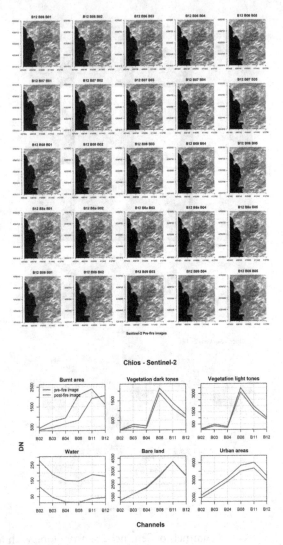

FIGURE 1.2 (Continued) Spectral channels of Sentinel-2 sensor along with few RGB color composites (left upper panel), RGB color composites from SWIR-NIR-Visible channels (right upper panel), histogram data plots of burned (post-fire image) vs. non-burned samples (pre-fire image) (left lower panel), and spectral signatures plots of various land cover types in the pre- and post-fire Sentinel-2 image (right lower panel). The two lower graphs concern only the B02, B03, B04, B08, B11, and B12 channels.

patterns have been observed in several studies and are almost identical with the spectral behavior of burned areas in similar environments (Koutsias and Karteris 1998, 2000; Koutsias, Mallinis, and Karteris 2009; Koutsias et al. 2013). More specifically, in the visible part of the electromagnetic spectrum the spectral signal of burned surfaces is similar to the spectral signal of vegetation in the pre-fire satellite image. Different studies carried out in the visible region indicated a varied behavior of burned areas depending on the type of affected vegetation, severity of burning, etc. (Arino et al. 2001; Pereira et al. 1997). More recently, Amos, Petropoulos, and Ferentinos (2019) concluded that the visible part of the electromagnetic spectrum was not well suited to discern burned from unburned land cover while the NBRb12 (Normalized Burn Ratio) (shortwave infrared 2: SWIR 2) produced the best results for detecting burned areas.

In addition to the visible part of the electromagnetic spectrum, a strong decrease in reflectance of burned surfaces is observed in the near-infrared part of the electromagnetic spectrum. The destruction of leaf cell structure, which reflects large quantities of incident solar radiation, is responsible for the reduction of the spectral signal of burned surfaces. Therefore, burned areas are expected to have lower reflectance in near-infrared bands than those of healthy vegetation. Many studies verify this spectral behavior of burned surfaces. In the near-infrared region the reflectance of burned surfaces is significantly lower than that of non-burned vegetation (Arino et al. 2001; Pereira et al. 1997). In contrast to near infrared, an increase in reflectance of burned surfaces is observed in the shortwave infrared that corresponds to Landsat TM or ETM+ band 7 or Sentinel-2 band 12. Burned areas are expected to have higher reflectance in shortwave infrared bands than those of healthy vegetation, due to the replacement of the vegetation layer with charcoal that reduces the water content which absorbs the radiation in this spectral region. Additionally, the spectral signal of burned surfaces in the SWIR region of the electromagnetic spectrum, that corresponds to Landsat TM or

ETM+ band 5 or Sentinel-2 band 11, occasionally and depending on the vegetation affected may be the same between the pre- and post-fire images or may decrease especially if the fire-affected vegetation is dry.

1.2 LANDSAT ENVIRONMENTAL APPLICATIONS: RECONSTRUCTION OF FIRE HISTORY

There is a long history of mapping burned areas using satellite images of multiple resolutions either at research level to develop and/or improve the methods to map them or to create operational products at global scale (Chuvieco et al. 2019). Maps of fire occurrence can improve our understanding for the protection and restoration of fire-affected natural ecosystems worldwide. Such maps depicting the spatially explicit fire history of an area, including variables such as fire frequency and fire return interval, are important tools to better understand the processes associated with wildfires, to assess the impacts of wildland fires on landscape dynamics, and to take decisions for the proper management of wildfires (Pleniou et al. 2012; Nioti, Dimopoulos, and Koutsias 2011).

Remote sensing has been an ideal tool to create maps of fire scars by collecting and processing the required data, as it covers large spatial and temporal extents within a cost- and time-efficient framework (Koutsias et al. 1999; Dalezios et al. 2017). The application of satellite data to burned area mapping has a long history in remote sensing studies (Richards 1984), and it is still an active research topic employing advanced techniques that integrate geo-statistics, support vector machines, and artificial neural networks (Lanorte et al. 2013; Gómez and Martín 2011; Petropoulos, Kontoes, and Keramitsoglou 2011; Boschetti, Stroppiana, and Brivio 2010; Mallinis and Koutsias 2012).

At a global scale fire products, such as those based on MODerate resolution Imaging Spectroradiometer (MODIS) (Justice et al. 2002), offer a unique data set; however, they are mostly devoted to continental scale studies such as characterizing global fire regimes (Chuvieco, Giglio, and Justice 2008) or estimating global biomass burning emissions (Korontzi et al. 2004). In the meantime, such systematic fire products are not common at local scales, mainly due to cost constraints on gathering and processing medium or high resolution satellite data series that until recently were not freely available at no cost. At a local scale, maps of fire occurrence are also important to scientists and fire managers for understanding the reasons of fire spread and ignition (Koutsias and Karteris 1998), investigating post-fire vegetation regeneration (Mitchell and Yuan 2010; Nioti et al. 2015), assessing land degradation (Bajocco, Salvati, and Ricotta 2011), avoiding mistakes when applying environmental policy and fire management (Kalabokidis et al. 2007), and providing consistent fire statistics (Loepfe, Lloret, and Román-Cuesta 2012).

Currently, U.S. Geological Survey (USGS) archived Landsat images are freely available to the public from the USGS Earth Resources Observation and Science (EROS) Center (http://glovis. usgs.gov/), and also ESA delivers worldwide Sentinel-2 satellite images free of charge from Copernicus Open Access Hub (https:// scihub.copernicus.eu/dhus/#/home). These historical archives cover large spatial and temporal extents at continental scale (Gutman and Masek 2012) and provide a unique opportunity to overcome cost constraints when reconstructing fire history globally at low to high spatial resolution. However, one problem that exists is how to process successfully, with high accuracy and without human interference, the chain of satellite images consisting of thousands of images. The training phase of the algorithm is a time-consuming procedure that if omitted would be a big advantage of the approach by saving processing time and cost. This is a critical issue in cases where many satellite images need to be analysed, such as for the spatially explicit reconstruction of recent fire history, where thousands of images might be used in the processing chain.

The development of completely automatic techniques to successfully map the burned areas is a challenge. Currently, there are efforts to develop automatic or semi-automatic techniques using medium resolution satellite images such as those of Landsat satellites (Koutsias et al. 2013; Bastarrika, Chuvieco, and Martín 2011; Bastarrika et al. 2014; Stroppiana et al. 2012; Boschetti et al. 2015), and also recently there is a development of Sentinel-2 burned area algorithms (Roteta et al. 2019; Roy et al. 2019). USGS delivered to the users Landsat Level-3 Burned Area (BA) that contains two acquisition-based raster data products that represent burn classification and burn probability using Surface Reflectance data from the U.S. Landsat Analysis Ready Data (ARD) (Hawbaker et al. 2017).

Koutsias et al. (2013) developed a semi-automatic method to map burned areas using multi-temporal acquisitions of Landsat satellite data, considering a pre- and post-fire image as a pair of images that could minimize spectral confusion with unburned surfaces. The proposed method consisted of a set of rules that are valid especially when the post-fire image was captured shortly after the fire event when the signal had its highest separability from the surrounding areas (Stroppiana et al. 2012). However, the rule-based approach is not free of errors (e.g. omission or commission), which eventually creates limitations to adopt this method for reconstructing the fire history in a fully automated mode. There is an effort to further improve the method by revisiting and improving the rules that have been developed in the first paper (Koutsias et al. 2013) especially for reducing the omission errors and capturing one special type of fires that occurred in dry-vegetated areas that the former approach was not capturing (Koutsias and Pleniou under prepararaion).

The rule-based approach consists of a set of five rules developed based on spectral properties of burned areas as compared to the pre-fire unburned vegetation and to the spectral signatures

TABLE 1.3   The rule-based approach consists of five rules that make use of either specific spectral bands only from the post-fire satellite image (rule1–rule3) or specific spectral bands from the pre- and post-fire satellite images (rule4–rule5)

Rule1

If Landsat *TM4 post* < Landsat *TM5 post then 1 (burned) otherwise 0 (unburned)*

Rule2

If Landsat TM5 post < Landsat *TM7 post + a Landsat TM7 post then 1 (burned) otherwise 0 (unburned) a = 0.35*

Rule3

If Landsat TM5 post > Landsat *TM1 post then 1 (burned) otherwise 0 (unburned)*

Rule4

If TM4 post + a TM4 post < *TM4 pre then 1 (burned) otherwise 0 (unburned) a = 0.2*

Rule5

If TM7 post > *TM7 pre + a TM7 pre or If TM5 post+ a TM5 post* < *TM5 pre then 1 (burned) otherwise 0 (unburned) a = 0.05*

---

of other land cover types found in post-fire satellite scenes. The spectral properties based on which the rules have been developed are presented in two graphs, one that corresponds to spectral signatures plots (Figure 1.2) and the second that corresponds to the histogram data plots (Figure 1.2). Based on these spectral patterns the arithmetic expressions of the developed five rules are presented in Table 1.3.

The R-code to implement the set of rules is presented in Table 1.4. An example of the results printed in R is illustrated in Figure 1.3 applied to both Landsat and Sentinel-2 images. Although this part refers to Landsat images, the whole procedure can be easily adapted to Sentinel-2 since there is a consistency of the spectral information between the two sensors.

TABLE 1.4   An example of R-code to implement the rule-based approach

"Landsat_pre", "Landsat_post" is the pre- and post-fire Landsat images. "name" is the name of USGS Landsat images (e.g. LT05_L1GS_193031_19850322_20171213_01_T2), "sr_band" corresponds to Landsat level-2 data.

```
library(raster)

if (substr(name, 3, 4)=="08") {channels<-c("2", "3", "4", "5", "6", "7")
} else { channels<-c("1", "2", "3", "4", "5", "7")}

Landsat_pre<-stack(paste(mypath, "\\",name,"\\",name,"_sr_band",channels,".tif", sep=""))

# Similar for Landsat_post

# Apply each rule, 1 when rule is true and 0 when rule is false
rule1<-raster(Landsat_post, 4)-raster(Landsat_post, 5)<0
rule2<-raster(Landsat_post, 5)-(raster(Landsat_post, 6)+0.35*raster(Landsat_post, 6))<0
rule3<-raster(Landsat_post, 5)-raster(Landsat_post, 1)>0
rule4<-raster(Landsat_post, 4)-(raster(Landsat_pre, 4)-0.2*raster(Landsat_pre, 4))<0
rule5a<-raster(Landsat_post, 6)-(raster(Landsat_pre, 6)+0.05*raster(Landsat_pre, 6))>0
rule5b<-raster(Landsat_post, 5)-(raster(Landsat_pre, 5)-0.05*raster(Landsat_pre, 5))<0
rule5<-rule5a+rule5b
rule5<-reclassify(rule5, c(1, 2, 1))

# Apply all rules, 1 when rule is true and 0 when rule is false
rules<-rule1*rule2* rule3* rule4* rule5*

# Vectorize the raster data to create vectors of only the pixels where all rules are true (1)
rules_p_2_a <- rasterToPolygons(rules_rules, fun=function(x){x==1}, dissolve=FALSE)
```

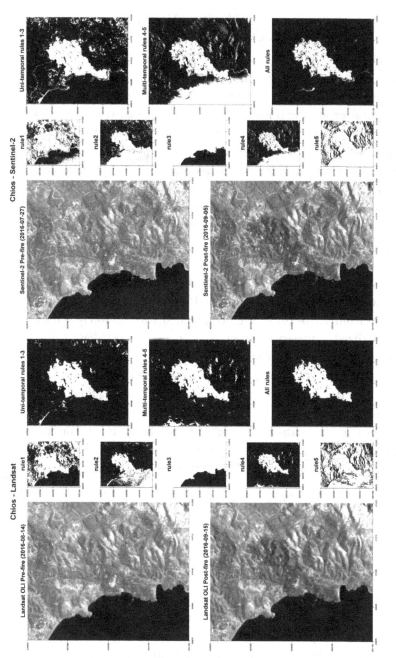

FIGURE 1.3   Illustrated results from the application of the rule-based approach to a study case in Chios, Greece, using both Landsat data (left panel) and Sentinel-2 data (right panel).

1.3 SENTINEL-2 ENVIRONMENTAL APPLICATIONS: MONITORING OF FIRE-AFFECTED AREAS

Vegetation phenology is an important element of vegetation characteristics that can be useful in vegetation monitoring especially when satellite remote sensing observations are used. In that sense temporal profiles extracted from spectral signal of time series Sentinel-2 or Landsat satellite images can be used to characterize vegetation phenology and thus are helpful for monitoring vegetation recovery in fire-affected areas. Studies of phenological patterns from satellite data are increasingly used nowadays especially under the perspective of freely available satellite imagery acquired with high temporal frequency at fine spatial resolution (Vrieling et al. 2018). Apart from the use of the original spectral bands, vegetation indices can be estimated and used to describe the monitoring patterns of fire-affected areas. For example, the spatio-temporal distribution of wildfires hotspots in Sardinia (Italy) from 2000–2013 have been related with remotely sensed Normalized Difference Vegetation Index (NDVI)-based fuel phenology patterns (Bajocco, Koutsias, and Ricotta 2017). A Vegetation Index (VI) is usually a linear transformation of two or more spectral bands aiming to enhance the spectral signal of the original spectral channels by creating a new spectral space which is sensitive to variations of vegetation attributes as for example leaf area index (LAI), percent green cover, chlorophyll content, green biomass, and absorbed photosynthetically active radiation (APAR).

It has been concluded that the original spectral channels, on which these indices are estimated, are sensitive to external vegetation parameters such as the spectral reflectance of the background soil (Pleniou and Koutsias 2013). In such cases, the influence of the soil in the reflectance values is different in the various spectral regions depending on its type. The use of such indices is also justified according to a recent study on the sensitivity of spectral reflectance values to different burn and vegetation ratios, based on which the Near Infrared (NIR) and

Shortwave Infrared (SWIR) are the most important channels to estimate the percentage of burned area, whereas the NIR and Red channels are the most important to estimate the percentage of vegetation in fire-affected areas. Additionally, it has been found that semi-burned classes are spectrally more consistent to their different fractions of scorched and non-scorched vegetation than the original spectral channels based on which these indices are estimated. Some of the indices that are considered typical and have been extensively applied for vegetation studies or in remote sensing of wildland fires are given in the following, although it is beyond our purpose to make an extensive evaluation concerning the phenology of fire-affected areas based on the vegetation indices.

Red-NIR Spectral Vegetation Indices
Normalized Difference Vegetation Index
NDVI (Rouse et al. 1973; Kriegler et al. 1969) is one of the most used and well-known vegetation indexes; it is sensitive to live green plants in multispectral remote sensing data. Conceptually, the index is based on the absorption of visible light (from 0.4 to 0.7 μm) (Red: MODIS band 1) from the pigment in plant leaves, chlorophyll, for the photosynthesis and the reflection of near-infrared light (from 0.7 to 1.1 μm) (NIR: MODIS band 2) because of the cell structure of the leaves. The index varies between −1 and +1. The formula to calculate NDVI from remote sensing spectral data is as follows:

$$NDVI = (NIR - Red)/(NIR + Red) \tag{1.1}$$

Ratio Vegetation Index (RVI)
RVI (Jordan 1969) is a simple vegetation index which simply divides the Near Infrared (NIR) by the visible spectral values (Red). Both RVI and NDVI basically measure the slope of the line between the origin of Red-NIR space and the Red-NIR value of the image pixel. The index values above +1. The

formula to calculate NDVI from remote sensing spectral data is as follows:

$$RVI = NIR/Red \qquad (1.2)$$

SWIR-NIR Spectral Indices

There is a series of indices that have been created by replacing the Red with the SWIR region of the electromagnetic spectrum. For example, some of these indices that have been proposed are Normalized Burn Ratio (NBR), Normalized Difference Water Index (NDWI), and Shortwave Infrared Water Stress Index (SIWSI).

Normalized Burn Ratio (NBR)

There is a particular spectral behavior of burned areas in the NIR and SWIR regions of the electromagnetic spectrum as described in Koutsias and Karteris (1998, 2000) that is the basis for the successful use of the NBR index (Key and Benson 1999, 2006), a modification of NDVI by replacing Red with SWIR, originally though proposed by Lopez-Garcia and Caselles (1991), who underlined the post-fire radiometric changes occurring in the SWIR, later verified by Koutsias and Karteris (2000). The replacement of the Red channel with the SWIR channels, that are sensitive to leaf water content because of absorption of the electromagnetic energy in this wavelength, has a long history in remote sensing (Ji et al. 2011). In fire mapping , NBR, which attempts to maximize reflectance change due to fire (Lozano, Suárez-Seoane, and de Luis 2007), has been found to be very successful in many studies. The formula to calculate NDVI from remote sensing spectral data is as follows:

$$NBR = (NIR - SWIR)/(NIR + SWIR) \qquad (1.3)$$

Normalized Difference Water Index (NDWI) or Shortwave Infrared Water Stress Index (SIWSI)

NDWI proposed by (Gao 1996) or SIWSI is a spectral vegetation index sensitive to water content of plant leaves (Fensholt

and Sandholt 2003). NDWI is calculated using the spectral channels that correspond to Near Infrared (NIR) and the Shortwave
Infrared (SWIR). The formula to calculate NDWI from remote
sensing spectral data is as follows:

$$NDWI = (NIR - SWIR)/(NIR + SWIR) \qquad (1.4)$$

Sentinel-2 data are provided by Copernicus Open Access Hub
(https://scihub.copernicus.eu/dhus/#/home) either as Level-
1C (S2MSI1C) Top-of-Atmosphere reflectances or as Level-2A
(S2MSI2A) Bottom-of-Atmosphere reflectances which are ~600
and ~800 MB respectively in size and 100 km × 100 km. Level-1C
can be transformed to Level-2A by using the Sen2Cor software
which is a processor for Sentinel-2 Level-2A product generation
and formatting developed by ESA. The software performs the
atmospheric, terrain, and cirrus correction of Top-of-Atmosphere
Level-1C input data and creates Bottom-of-Atmosphere, option-
ally terrain and cirrus corrected reflectance images. Additionally,
it creates Aerosol Optical Thickness, Water Vapor, Scene Classification Maps, and Quality Indicators for cloud and snow proba-
bilities (https://step.esa.int/main/third-party-plugins-2/sen2cor/).
The image data are provided in GML-JPEG2000 format. To read
them in R (R Core Team 2017) the "rgdal" package (Bivand, Keitt,
and Rowlingson 2018) can be used along with the "raster" package (Hijmans 2019). An example of the R-script is provided in
Table 1.5.

The first step for creating temporal profiles of the spectral signal
of time series Sentinel-2 or Landsat satellite images is to extract
the spectral information from all available time series satellite
images. An example of the R-script to read all time series satellite
images, given as separate TIF files, is provided in Table 1.6.

One of the problems arising when extracting the spectral values of specific pixels from all available Sentinel-2 satellite images
is the existence of noise that can be created from various sources
like clouds, cloud shadows, etc. For the case of Sentinel-2, the

TABLE 1.5 An example of R-code to read Sentinel-2 image data of GML-JPEG2000 format

```
library(rgdal)
library(raster)

# Read 10m Bottom-of-Atmosphere reflectances Sentinel bands using a loop,
indicated by i. Here an example for band B02
list_dirs<-grep(list.dirs(path=mypath, full.names = FALSE, recursive = FALSE),
pattern=c("L2A_"), inv=F, value=T)
for(i in 1:length(list_dirs)){
name_dir<-list.files(paste(mypath, "\\", list_dirs[i], "\\GRANULE\\", sep=""))
name_file_ext<-list.files(paste(mypath, "\\", list_dirs[i], "\\GRANULE\\",
name_dir, "\\IMG_DATA\\R10m\\", sep=""), pattern=c("B0"))

B02<-readGDAL(paste(mypath, "\\", list_dirs[i], "\\GRANULE\\", name_dir, "\\
IMG_DATA\\R10m\\", grep(name_file_ext, pattern=c("B02"), inv=F, value=T),
sep=""))

B02<-stack(B02)
# Similar to other Sentinel-2 bands
}
```

TABLE 1.6 An example of the R-script to read all time series satellite images given as separate TIF files using either non-parallel processing or parallel processing

```
library(rgdal)
library(raster)
library(foreach)
library(doParallel)

shape<-readOGR(paste(path_input, files[f], sep=""))
shape<-SpatialPoints(data.frame(xPoints = shape@coords[,c(1)], yPoints =
shape@coords[,c(2)]))
extent<-extent(shape)+c(-100, +100, -100, +100)

list_tifs<-gsub(".tif", "", list.files(paste(path_data, sep=""), pattern=".tif"))

tmp <- brick(paste(path_data, list_tifs[1], ".tif", sep=""))
extract<-extract(crop(tmp, extent), shape)
rows<-nrow(extract)
df<-data.frame(matrix(ncol = rows*6+1, nrow = 0))
for(i in 1:rows){
colnames(df)[((i-1)*6+1+1):((i-1)*6+6+1)] <- paste(c("B02", "B03", "B04",
"B08", "B11", "B12"), paste("p",i,sep = ""), sep = "_")
}
colnames(df)[1]<-"Satname"
```

(*Continued*)

TABLE 1.6   (Continued) An example of the R-script to read all time series satellite images given as separate TIF files using either non-parallel processing or parallel processing

```
# non-parallel processing
for (i in 1:length(list_tifs)) {
r1 <- brick(paste(path_data, list_tifs[i], ".tif", sep=""), extent)
extract<-extract(r1, shape)

for(j in 1:rows){
df[i,((j-1)*6+1+1):((j-1)*6+6+1)]<-extract[j,]
}
}
df[,1]<-paste(list_tifs)

# parallel processing
cores=detectCores()
cl <- makeCluster(cores[1]-1)
registerDoParallel(cl)
finalMatrix = foreach(i=1:length(list_tifs), .combine=rbind) %dopar% {
library(raster)
tempMatrix = {}
tempMatrix = extract(brick(paste(path_data, list_tifs[i], ".tif", sep=""), extent), shape)
}
stopCluster(cl)

j=1
for(i in 1:length(list_tifs)){
for(j in 1:rows){
df[i,((j-1)*6+1+1):((j-1)*6+6+1)]<-finalMatrix[(i-1)*rows+j,]
}
}
df[,1]<-paste(list_tifs)
```

Sen2Cor's Scene Classification output can be used to discard any pixels classified as saturated or defective, cloud shadow, medium to high cloud probability, thin cirrus, or snow/ice (Vrieling et al. 2018). Additionally to this, processing of the time series can identify anomalies created from such processes and remove them from from the original data (Hirschmugl et al. 2017).

Vrieling et al. (2018) showed that phenological parameters can be retrieved from Sentinel-2 image time series although large differences existed between the satellite-derived NDVI series with camera-derived greenness chromatic coordinate (GCC). In their work, they applied the phenology retrieval approach to NDVI series for all pixels for mapping spatial patterns of phenology demonstrating therefore the potential of the Sentinel-2 mission for providing spatially detailed retrievals of phenology. Short-term phenospectral dynamics like spectral changes indicated by specific spectral features as a function of phenology were assessed in Arroyo-Mora et al. (2018). They found that Sentinel-2A reflectance and vegetation indices products capture the short-term phenospectral changes at the landscape level and are consistent with the hyperspectral imagery (HSI) resampled products. The seasonal dynamics of leaf area index (LAI) and aboveground biomass (AGB) were estimated using Sentinel-1 (S1), Landsat-8 (LC8), and Sentinel-2 (S2) data by Wang et al. (2019). Their results indicated that intergraded Landsat 8 and Sentinel-2 data can be enough to capture the seasonal dynamics of grasslands at a 10- to 30-m spatial resolution. Automatic methods, based on crop phenology, are also being developed to detect, for example, rice using the time series of Sentinel-2 imagery (Rad et al. 2019).

Techniques to analyze the time series data are coming either from time series statistics or from other domains like the dynamic time warping (DTW) approach, which, according to Belgiu and Csillik (2018), originally was developed for speech recognition (Sakoe and Chiba 1978) and later introduced in the analysis of time series data (Maus et al. 2019; Petitjean, Inglada, and Gancarski 2012). The dynamic time warping algorithm measures the similarity between two temporal sequences that may vary in speed. The method calculates an optimal match between two given sequences (e.g. time series) with certain restriction and rules. The method has been applied in crop mapping to consider the seasonality of crops an issue similar to the monitoring of fire-affected areas because of the seasonality of vegetation which can vary depending of course on the type of vegetation (Csillik et al. 2019). It has been also applied to improve the spatial and temporal resolution of remote sensing snow products (Berman et al. 2018).

Finally, for the time series of the burned case we subtracted the value to the position on time t with the corresponding value on t-1, and in this case there is an excellent removal of all other signals except the signal corresponded to the date of fire occurrence of fire occurrence (Figure 1.4).

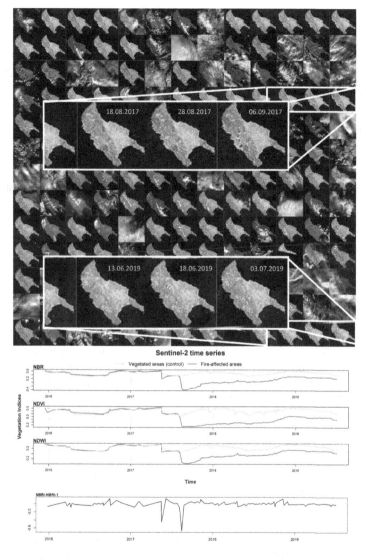

FIGURE 1.4 All available Sentinel-2 satellite images of Zakynthos island in Greece where some fires occurred during the summer of 2017. Time series of spectral values of the pixels corresponding to those fires (red lines) and to unscorched vegetation around the fires (green lines) have been extracted, and vegetation indices have been estimated to monitor fire recovery. The burned area can be identified in these time series by subtracting the value to the position on time t with the corresponding value on t-1.

Acknowledgments

The Landsat satellite images were freely obtained from the U.S. Geological Survey (www.usgs.gov). Sentinel-2 satellite images were freely obtained from the ESA Copernicus Open Access Hub (https://scihub.copernicus.eu/dhus/#/home).

REFERENCES

Amos, C., G. P. Petropoulos, and K. P. Ferentinos. 2019. "Determining the Use of Sentinel-2A MSI for Wildfire Burning & Severity Detection." *International Journal of Remote Sensing* 40 (3): 905–930. https://doi.org/10.1080/01431161.2018.1519284.

Arino, O., I. Piccolini, E. Kasischke, F. Siegert, E. Chuvieco, P. Martin, Z. Li, R. Fraser, H. Eva, D. Stroppiana, J. Pereira, J. M. N. Silva, D. Roy, and P. M. Barbosa. 2001. "Methods of Mapping Burned Surfaces in Vegetation Fires." In *Global and Regional Vegetation Fire Monitoring from Space: Planning a Coordinated International Effort*, edited by F. J. Ahern, J. G. Goldammer and C. O. Justice, 227–255. The Hague, The Netherlands: SPB, Academic Publishing.

Arroyo-Mora, J. P., M. Kalacska, R. Soffer, G. Ifimov, G. Leblanc, E. S. Schaaf, and O. Lucanus. 2018. "Evaluation of Phenospectral Dynamics With Sentinel-2A Using a Bottom-Up Approach in a Northern Ombrotrophic Peatland." *Remote Sensing of Environment* 216: 544–560. https://doi.org/10.1016/j.rse.2018.07.021.

Bajocco, Sofia, Nikos Koutsias, and Carlo Ricotta. 2017. "Linking Fire Ignitions Hotspots and Fuel Phenology: The Importance of Being Seasonal." *Ecological Indicators* 82: 433–440. https://doi.org/10.1016/j.ecolind.2017.07.027.

Bajocco, Sofia, L. Salvati, and Carlo Ricotta. 2011. "Land Degradation Versus Fire: A Spiral Process?" *Progress in Physical Geography* 35 (1): 3–18. https://doi.org/10.1177/0309133310380768.

Bastarrika, Aitor, Maite Alvarado, Karmele Artano, Maria Martinez, Amaia Mesanza, Leyre Torre, Rubén Ramo, and Emilio Chuvieco. 2014. "BAMS: A Tool for Supervised Burned Area Mapping Using Landsat Data." *Remote Sensing* 6 (12): 12360–12380.

Bastarrika, Aitor, E. Chuvieco, and M. P. Martín. 2011. "Mapping Burned Areas From Landsat TM/ETM+ Data With a Two-Phase Algorithm: Balancing Omission and Commission Errors." *Remote Sensing of Environment* 115 (4): 1003–1012Belgiu, Mariana, and Ovidiu Csillik. 2018. "Sentinel-2 Cropland Mapping Using Pixel-

Based and Object-Based Time-Weighted Dynamic Time Warping Analysis." *Remote Sensing of Environment* 204: 509–523. https://doi.org/10.1016/j.rse.2017.10.005.

Berman, E. E., D. K. Bolton, N. C. Coops, Z. K. Mityok, G. B. Stenhouse, and R. D. D. Moore. 2018. "Daily Estimates of Landsat Fractional Snow Cover Driven by MODIS and Dynamic Time-Warping." *Remote Sensing of Environment* 216: 635–646. https://doi.org/10.1016/j.rse.2018.07.029.

Bivand, Roger, Tim Keitt, and Barry Rowlingson. 2018. *rgdal: Bindings for the 'Geospatial' Data Abstraction Library.* https://CRAN.R-project.org/package=rgdal.

Boschetti, Luigi, David P. Roy, Christopher O. Justice, and Michael L. Humber. 2015. "MODIS – Landsat Fusion for Large Area 30 m Burned Area Mapping." *Remote Sensing of Environment* 161: 27–42. https://doi.org/10.1016/j.rse.2015.01.022.

Boschetti, M., D. Stroppiana, and P. A. Brivio. 2010. "Mapping Burned Areas in a Mediterranean Environment Using Soft Integration of Spectral Indices from High-Resolution Satellite Images." *Earth Interactions* 14: 1–20. https://doi.org/10.1175/2010EI349.1.

Chuvieco, Emilio, L. Giglio, and C. Justice. 2008. "Global Characterization of Fire Activity: Toward Defining Fire Regimes From Earth Observation Data." *Global Change Biology* 14 (7): 1488–1502.

Chuvieco, Emilio, Florent Mouillot, Guido R. van der Werf, Jesús San Miguel, Mihai Tanasse, Nikos Koutsias, Mariano García, Marta Yebra, Marc Padilla, Ioannis Gitas, Angelika Heil, Todd J. Hawbaker, and Louis Giglio. 2019. "Historical Background and Current Developments for Mapping Burned Area from Satellite Earth Observation." *Remote Sensing of Environment* 225: 45–64. https://doi.org/10.1016/j.rse.2019.02.013.

Csillik, O., M. Belgiu, G. P. Asner, and M. Kelly. 2019. "Object-Based Time-Constrained Dynamic Time Warping Classification of Crops Using Sentinel-2." *Remote Sensing* 11 (10). https://doi.org/10.3390/rs11101257.

Dalezios, N. R., K. Kalabokidis, N. Koutsias, and C. Vasilakos. 2017. "Wildfires and Remote Sensing: An Overview." In: Petropoulos, G.P., Islam, T. (Eds.), *Remote Sensing of Hydrometeorological Hazards.* CRC Press, pp. 211–236.

Fensholt, Rasmus, and Inge Sandholt. 2003. "Derivation of a Shortwave Infrared Water Stress Index From MODIS Near- and Shortwave Infrared Data in a Semiarid Environment." *Remote Sensing of Environment* 87 (1): 111–121. https://doi.org/10.1016/j.rse.2003.07.002.

Fernández-Manso, A., O. Fernández-Manso, and C. Quintano. 2016. "SEN-TINEL-2A Red-Edge Spectral Indices Suitability for Discriminating Burn Severity." *International Journal of Applied Earth Observation and Geoinformation* 50: 170–175. https://doi.org/10.1016/j.jag.2016.03.005.

Gao, Bo-cai. 1996. "NDWI – A Normalized Difference Water Index for Remote Sensing of Vegetation Liquid Water From Space." *Remote Sensing of Environment* 58 (3): 257–266. https://doi.org/10.1016/S0034-4257(96)00067-3.

Gómez, Israel, and M. Pilar Martín. 2011. "Prototyping an Artificial Neural Network for Burned Area Mapping on a Regional Scale in Mediterranean Areas Using MODIS Images." *International Journal of Applied Earth Observation and Geoinformation* 13 (5): 741–752.

Gutman, Garik, and Jeffrey G. Masek. 2012. "Long-Term Time Series of the Earth's Land-Surface Observations From Space." *International Journal of Remote Sensing* 33 (15): 4700–4719. https://doi.org/10.1080/01431161.2011.638341.

Hawbaker, Todd J., Melanie K. Vanderhoof, Yen-Ju Beal, Joshua D. Takacs, Gail L. Schmidt, Jeff T. Falgout, Brad Williams, Nicole M. Fairaux, Megan K. Caldwell, Joshua J. Picotte, Stephen M. Howard, Susan Stitt, and John L. Dwyer. 2017. "Mapping Burned Areas Using Dense Time-Series of Landsat Data." *Remote Sensing of Environment* 198: 504–522. https://doi.org/10.1016/j.rse.2017.06.027.

Hijmans, Robert J. 2019. "raster: Geographic Data Analysis and Modeling." *R package version 2.8-19*. https://CRAN.R-project.org/package=raster.

Hirschmugl, Manuela, Heinz Gallaun, Matthias Dees, Pawan Datta, Janik Deutscher, Nikos Koutsias, and Mathias Schardt. 2017. "Methods for Mapping Forest Disturbance and Degradation From Optical Earth Observation Data: A Review." *Current Forestry Reports* 3 (1): 32–45. https://doi.org/10.1007/s40725-017-0047-2.

Huang, H., D. P. Roy, L. Boschetti, H. K. Zhang, L. Yan, S. S. Kumar, J. Gomez-Dans, and J. Li. 2016. "Separability Analysis of Sentinel-2A Multi-Spectral Instrument (MSI) Data for Burned Area Discrimination." *Remote Sensing* 8 (10). https://doi.org/10.3390/rs8100873.

Ji, Lei, Li Zhang, Bruce K. Wylie, and Jennifer Rover. 2011. "On the Terminology of the Spectral Vegetation Index (NIR-SWIR)/(NIR+SWIR)." *International Journal of Remote Sensing* 32 (21): 6901–6909. https://doi.org/10.1080/01431161.2010.510811.

Jordan, Carl F. 1969. "Derivation of Leaf-Area Index From Quality of Light on the Forest Floor." *Ecology* 50 (4): 663–666. https://doi.org/10.2307/1936256.

Justice, C. O., L. Giglio, S. Korontzi, J. Owens, J. T. Morisette, D. Roy, J. Descloitres, S. Alleaume, F. Petitcolin, and Y. Kaufman. 2002. "The MODIS Fire Products." *Remote Sensing of Environment* 83 (1–2): 244–262.

Kalabokidis, K. D., N. Koutsias, P. Konstantinidis, and C. Vasilakos. 2007. "Multivariate Analysis of Landscape Wildfire Dynamics in a Mediterranean Ecosystem of Greece." *Area* 39 (3): 392–402.

Key, C. H., and N. C. Benson. 1999. "Measuring and Remote Sensing of Burn Severity: The CBI and NBR." Joint Fire Science Conference and Workshop, Boise.

Key, C. H., and N. C. Benson. 2006. "Landscape Assessment: Ground Measure of Severity, the Composite Burn Index; and Remote Sensing of Severity, the Normalized Burn Ratio." In *FIREMON: Fire Effects Monitoring and Inventory System*, edited by D. C. Lutes, R. E. Keane, J. F. Caratti, C. H. Key, N. C. Benson, S. Sutherland and L. J. Gangi, 1–51. Ogden, UT: USDA Forest Service, Rocky Mountain Research Station.

Korontzi, S., D. P. Roy, C. O. Justice, and D. E. Ward. 2004. "Modeling and Sensitivity Analysis of Fire Emissions in Southern Africa During SAFARI 2000." *Remote Sensing of Environment* 92 (2): 255–275.

Koutsias, Nikos, and Michael Karteris. 1998. "Logistic Regression Modelling of Multitemporal Thematic Mapper Data for Burned Area Mapping." *International Journal of Remote Sensing* 19 (18): 3499–3514.

Koutsias, Nikos, and Michael Karteris. 2000. "Burned Area Mapping Using Logistic Regression Modeling of a Single Post-Fire Landsat-5 Thematic Mapper Image." *International Journal of Remote Sensing* 21 (4): 673–687. https://doi.org/10.1080/014311698213777.

Koutsias, Nikos, Michael Karteris, A. Fernandez-Palacios, C. Navarro, J. Jurado, R. Navarro, and A. Lobo. 1999. "Burned Land Mapping at Local Scale." In *Remote Sensing of Large Wildfires in the European Mediterranean Basin*, edited by E. Chuvieco, 157–187. Berlin Heidelberg: Springer-Verlag.

Koutsias, Nikos, G. Mallinis, and Michael Karteris. 2009. "A Forward/Backward Principal Component Analysis of Landsat-7 ETM+ Data to Enhance the Spectral Signal of Burnt Surfaces." *ISPRS Journal of Photogrammetry & Remote Sensing* 64 (1): 37–46.

Koutsias, Nikos, and M. Pleniou. under prepararation. "A Rule-Based Semi-Automatic Method to Map Burned Areas Using Landsat Images – Revisited and Improved."

Koutsias, Nikos, M. Pleniou, G. Mallinis, F. Nioti, and N. I. Sifakis. 2013. "A Rule-Based Semi-Automatic Method to Map Burned Areas: Exploring the USGS Historical Landsat Archives to Reconstruct Recent Fire History." *International Journal of Remote Sensing* 34 (20): 7049–7068.

Kriegler, F., W. Malila, R. Nalepka, and W. Richardson. 1969. "Preprocessing Transformations and their Effect on Multispectral Recognition." Proceedings of the 6th International Symposium on Remote Sensing of Environment, Ann Arbor, MI.

Lanorte, A., M. Danese, R. Lasaponara, and B. Murgante. 2013. "Multiscale Mapping of Burn Area and Severity Using Multisensor Satellite Data and Spatial Autocorrelation Analysis." *International Journal of Applied Earth Observation and Geoinformation* 20: 42–51.

Loepfe, Lasse, Francisco Lloret, and Rosa María Román-Cuesta. 2012. "Comparison of Burnt Area Estimates Derived From Satellite Products and National Statistics in Europe." *International Journal of Remote Sensing* 33 (12): 3653–3671. https://doi.org/10.1080/01431 161.2011.631950.

Lopez Garcia, M. J., and V. Caselles. 1991. "Mapping Burns and Natural Reforestation Using Thematic Mapper Data." *Geocarto International* 6 (1): 31–37.

Lozano, F. Javier, Susana Suárez-Seoane, and Estanislao de Luis. 2007. "Assessment of Several Spectral Indices Derived From Multi-Temporal Landsat Data for Fire Occurrence Probability Modelling." *Remote Sensing of Environment* 107 (4): 533–544.

Mallinis, G., and N. Koutsias. 2012. "Comparing Ten Classification Methods for Burned Area Mapping in a Mediterranean Environment Using Landsat TM Satellite Data." *International Journal of Remote Sensing* 33 (14): 4408–4433.

Maus, Victor, Gilberto Câmara, Marius Appel, and Edzer Pebesma. 2019. "dtwSat: Time-Weighted Dynamic Time Warping for Satellite Image Time Series Analysis in R." *Journal of Statistical Software* 1 (5).

Mitchell, M., and F. Yuan. 2010. "Assessing Forest Fire and Vegetation Recovery in the Black Hills, South Dakota." *GIScience and Remote Sensing* 47 (2): 276–299.

Nioti, Foula, P. Dimopoulos, and Nikos Koutsias. 2011. "Correcting the Fire Scar Perimeter of a 1983 Wildfire Using USGS Archived Landsat Satellite Data." *GIScience & Remote Sensing* 48 (4): 600–613.

Nioti, Foula, Fotios Xystrakis, Nikos Koutsias, and Panayotis Dimopoulos. 2015. "A Remote Sensing and GIS Approach to Study the

Long-Term Vegetation Recovery of a Fire-Affected Pine Forest in Southern Greece." *Remote Sensing* 7 (6): 7712–7731.

Pereira, J. M. C., Emilio Chuvieco, A. Beaudoin, and N. Desbois. 1997. "Remote Sensing of Burned Areas: A Review." In *A Review of Remote Sensing Methods for the Study of Large Wildland Fires*, edited by Emilio Chuvieco, 127–183. Alcala de Henares, Spain: Universidad de Alcala.

Petitjean, F., J. Inglada, and P. Gancarski. 2012. "Satellite Image Time Series Analysis Under Time Warping." *IEEE Transactions on Geoscience and Remote Sensing* 50 (8): 3081–3095. https://doi.org/10.1109/TGRS.2011.2179050.

Petropoulos, George P., Charalambos Kontoes, and Iphigenia Keramitsoglou. 2011. "Burnt Area Delineation From a Uni-Temporal Perspective Based on Landsat TM Imagery Classification Using Support Vector Machines." *International Journal of Applied Earth Observation and Geoinformation* 13 (1): 70–80.

Pleniou, Magdalini, and Nikos Koutsias. 2013. "Sensitivity of Spectral Reflectance Values to Different Burn and Vegetation Ratios: A Multi-Scale Approach Applied in a Fire Affected Area." *ISPRS Journal of Photogrammetry & Remote Sensing* 79: 199–210. https://doi.org/10.1016/j.isprsjprs.2013.02.016.

Pleniou, Magdalini, Fotios Xystrakis, Panayotis Dimopoulos, and Nikos Koutsias. 2012. "Maps of Fire Occurrence – Spatially Explicit Reconstruction of Recent Fire History Using Satellite Remote Sensing." *Journal of Maps* 8 (4): 499–506. https://doi.org/10.1080/17445647.2012.743866.Rad, A. M., D. Ashourloo, H. S. Shahrabi, and H. Nematollahi. 2019. "Developing an Automatic Phenology-Based Algorithm for Rice Detection Using Sentinel-2 Time-Series Data." *IEEE Journal of Selected Topics in Applied Earth Observations and Remote Sensing* 12 (5): 1471–1481. https://doi.org/10.1109/JSTARS.2019.2906684.

R Core Team. 2017. *R: A Language and Environment for Statistical Computing*. Vienna, Austria: R Core Team.

Richards, J. A. 1984. "Thematic Mapping From Multitemporal Image Data Using the Principal Component Transformation." *Remote Sensing of Environment* 16: 35–46.

Roteta, E., A. Bastarrika, M. Padilla, T. Storm, and E. Chuvieco. 2019. "Development of a Sentinel-2 Burned Area Algorithm: Generation of a Small Fire Database for Sub-Saharan Africa." *Remote Sensing of Environment* 222: 1–17. https://doi.org/10.1016/j.rse.2018.12.011.

Rouse, J. W., R. H. Haas, J. A. Schell, and D. W. Deering. 1973. "Monitoring Vegetation Systems in the Great Plains With ERTS." Third ERTS Symposium.

Roy, D. P., H. Huang, L. Boschetti, L. Giglio, L. Yan, H. H. Zhang, and Z. Li. 2019. "Landsat-8 and Sentinel-2 Burned Area Mapping – A Combined Sensor Multi-Temporal Change Detection Approach." *Remote Sensing of Environment* 231. https://doi.org/10.1016/j.rse.2019.111254.

Sakoe, H., and S. Chiba. 1978. "Dynamic Programming Algorithm Optimization for Spoken Word Recognition." *IEEE Transactions on Acoustics, Speech, and Signal Processing* 26 (1): 43–49. https://doi.org/10.1109/TASSP.1978.1163055.

Stroppiana, D., G. Bordogna, P. Carrara, M. Boschetti, L. Boschetti, and P. A. Brivio. 2012. "A Method for Extracting Burned Areas From Landsat TM/ETM+ Images by Soft Aggregation of Multiple Spectral Indices and a Region Growing Algorithm." *ISPRS Journal of Photogrammetry and Remote Sensing* 69: 88–102.

Vrieling, A., M. Meroni, R. Darvishzadeh, A. K. Skidmore, T. Wang, R. Zurita-Milla, K. Oosterbeek, B. O'Connor, and M. Paganini. 2018. "Vegetation Phenology From Sentinel-2 and Field Cameras for a Dutch Barrier Island." *Remote Sensing of Environment* 215: 517–529. https://doi.org/10.1016/j.rse.2018.03.014.

Wang, J., X. Xiao, R. Bajgain, P. Starks, J. Steiner, R. B. Doughty, and Q. Chang. 2019. "Estimating Leaf Area Index and Aboveground Biomass of Grazing Pastures Using Sentinel-1, Sentinel-2 and Landsat Images." *ISPRS Journal of Photogrammetry and Remote Sensing* 154: 189–201. https://doi.org/10.1016/j.isprsjprs.2019.06.007.

## CHAPTER 2

Environmental Applications of Medium Resolution Remote Sensing Imaging

Alexandra Gemitzi

2.1 INTRODUCTION: BACKGROUND INFORMATION OF MODIS PRODUCTS

MODerate resolution Imaging Spectroradiometer (MODIS) products are divided into three major categories based on the level of preprocessing and the target applications. Therefore, Level-1 products provide raw radiances along with cloud mask data and geolocation fields. Higher level MODIS land, atmosphere, cryosphere products are developed from the MODIS Adaptive Processing System (MODAPS) and distributed to the scientific community through three Active Archive Centers (DAACs) (https://modis.

gsfc.nasa.gov/data/dataprod/index.php). Ocean color and sea surface temperature MODIS products are supplied by the Ocean Color Data Processing System (OCDPS). All MODIS products have two primary sources of information, i.e. MODIS Aqua and MODIS Terra data sets. The main difference between those two data sources is related to the orbiting cycles of the two satellites. Terra circles a morning orbit passing from north to south over the equator in the morning while Aqua follows an afternoon orbit, i.e. passes from south to north over the equator in the afternoon. The difference in orbiting cycles causes differences in the acquired daily images due to different solar zenith angle but also different cloud cover conditions for daily MODIS images. Despite the primary data source and the associated differences, MODIS products use the same algorithms for various land products for both Aqua and Terra acquisitions. After processing the swath MODIS data, gridded products at daily, multi-day (8-day, 16-day, and monthly), and yearly time scales are provided. The spatial resolution of the girded products spams from 250 m to 5,600 m. Presently, under the latest MODIS collection, i.e. Collection 6, 152 science products from Aqua, Terra, or combined are freely available to the scientific community, concerning a wide variety of environmental parameters such as vegetation indices, Bidirectional Reflectance Distribution Function (BRDF)/Albedo, LST/Emissivity, burned areas, thermal anomalies and fires, land cover types, gross and net primary productivity, and net evapotranspiration.

Within the present chapter, three applications of widely used MODIS land products will be presented along with R-codes for data downloading and processing.

2.2 MODIS LST PRODUCTS: ESTIMATING LAND SURFACE TEMPERATURE TRENDS OVER GREECE

The terms "surface temperature" and "land surface temperature" (LST) are frequently used in scientific applications and refer to three different physical concepts with different physical principles of measurement. Therefore, surface temperature and LST may

refer to radiometric surface temperature (or skin temperature), surface air temperature (defined as air temperature measured at shelter
height), or aerodynamic temperature (also defined as temperature at
the height of roughness length for heat). The use of the same terminology for different physical conceptualizations causes ambiguities
and difficulties in data analysis especially in multidisciplinary environmental applications (Jin and Dickinson 2010). Considering that
there are no direct methods of measuring aerodynamic temperature,
this property is usually associated to LST (Chehbouni et al. 1996),
and consequently climate studies and assessments are composed
of analysis of either air temperature or LST data. In climate studies we should keep in mind that LST is the radiative skin temperature and is measured with infrared (IR) radiometers which measure
thermal radiance from land surface (Jin and Dickinson 2010) only
during clear sky days, as thermal infrared radiance does not penetrate clouds to reach satellites. The atmospheric attenuation effect
should be removed when measuring LST from satellite missions. It
is well known that the ozone of the atmosphere absorbs most of the
radiance from ground, but there is a specific spectral window where
there is minimal loss of ground radiance, i.e. the 10- to 13-mm range
(Qin et al. 2001). Therefore, this specific spectral window has been
selected for the thermal channels of MODIS sensors along with
other popular satellite platforms like Landsat Thematic Mapper and
National Oceanic and Atmospheric Administration-Advanced Very
High Resolution Radiometer (NOAA-AVHRR) (Mildrexler, Zhao,
and Running 2011). LST differs from air temperature. LST can be
regarded as the feeling of heat when touching Earth's surface. On
the other hand, air temperature is typically measured 1.5 m above
the ground level with several sensors, including thermistors, ther-
mocouples, and mercury thermometers protected from radiation
and adequately ventilated (Mildrexler, Zhao, and Running 2011).
LST is positively correlated to air temperature with the strength of
the relationship being dependent on the land cover type and cloud
conditions (Mildrexler, Zhao, and Running 2011; Gallo et al. 2011).
Mean differences between air temperature and LST are reported to

be less than 2 degrees C under cloudy conditions and higher than 2 degrees C under clear sky (Gallo et al. 2011). Mildrexler, Zhao, and Running (2011) indicate that as LST is tightly related to the radiative and thermodynamic characteristics of Earth's surface, it might be a more suitable variable to use in calculations of the global average temperature using the radiative-convective concept (Pielke et al. 2007). Scientific community requirements concerning accuracy of satellite-derived LST at a spatial resolution of 1- to 10-km range from 0.5 degrees C–2 degrees C (Wang, Liang, and Meyers 2008). Such accuracy constitutes the LST products useful and acceptable for meteorological, hydrological, and agricultural studies and assessments. MODIS LST products have been validated extensively in various research works (Coll et al. 2005; Wang, Liang, and Meyers 2008; Z Wan and Li 2008; Lu et al. 2018), and their accuracy is reported to 1 degrees C for surfaces with known emissivity (Z. Wan et al. 2002; Wang, Liang, and Meyers 2008). Considering that MODIS LST is provided since 2000, with both daytime and nighttime LST retrievals, their high accuracy, their free and easy accessibility, constitute MODIS LST products as state of the art output. This is the reason why the scientific community received with enthusiasm MODIS LST products which have nowadays become standard input for environmental assessments.

Data Access for MODIS LST Products

When dealing with climate research and use of remotely sensed environmental information, scientists need to decide on the most suitable product for their needs and find out how they can access and analyze data to have scientifically sound assessments. Collection 6 is the most recent among MODIS product collections and uses an improved algorithm which has been further validated compared to Collection 5. The MxD11 product is composed of one swath (MxD11_L2/) and six gridded products (MxD11A1, MxD11A2, MxD11B1, MxD11C1, MxD11C2, and MxD11C3) (Zhengming Wan 2013) (https://modis.gsfc.nasa.gov/data/dataprod/mod11.php).

An improvement to MxD11 products which have been documented to underestimate LST with a cold LST bias ranging from 3–5 K over the arid and semi-arid areas (Hulley et al. 2016) has been achieved in Collection 6 MxD21 products which are composed of two swath (Level-2) and six gridded products (Level-3). Level-2 is composed of the Land Surface Temperature/Emissivity Daily 5-min L2 Swath 1km products named as MOD21_L2 (Terra) and MYD21_L2 (Aqua). Level-3 comprises global 1km gridded products in sinusoidal projection, namely Land Surface Temperature/ Emissivity Daily L3 Global 1km SIN Grid Day (MOD21A1D/ MYD21A1D), Land Surface Temperature/Emissivity Daily L3 Global 1km SIN Grid Night (MOD21A1N/MYD21A1N), and Land Surface Temperature/Emissivity 8-day L3 Global 1 km (MOD21A2/MYD21A2) (https://modis.gsfc.nasa.gov/data/data-prod/mod21.php). However, presently the MxD21 LST products are released with provisional maturity status and shorter time period (from 2000 to 2005) due to technical problems in band 29 record after 2005. Therefore, within this book MxD11 products are used.

NASA has developed many tools and resources for data downloading and manipulating, including data access utilities, e-learning resources and webinars, webservices, and computer codes. Downloading utilities include the following applications: AppEEARS (https://lpdaac.usgs.gov/tools/appeears/) for search, subset, and decode quality; Data Pool (https://lpdaac.usgs.gov/ tools/data-pool/) for direct download; LDOPE (https://lpdaac. usgs.gov/tools/ldope/) for decoding quality; NASA Earthdata Search (https://lpdaac.usgs.gov/tools/earthdata-search/) for search, subset, order, direct download, and browse image preview; DAAC2Disk Utility (https://lpdaac.usgs.gov/tools/daac2 diskscripts/) for direct download; USGS EarthExplorer (https:// lpdaac.usgs.gov/tools/usgs-earthexplorer/) for search, direct download, and browse image preview. Whichever users select, it is strongly recommended that they should first refer to the provided user's guide for each specific MODIS product, provided by

NASA's scientific teams. A handy application for subsetting and downloading MODIS land products along with R-codes for data processing and visualization is the Global Subsets Tool (https://modis.ornl.gov/cgi-bin/MODIS/global/subset.pl). A unique capability of this application is that even inexperienced MODIS data users may subset and access filtered and scaled data and use various spatial and temporal statistical indices for many environmental variables.

Many users, however, prefer to write their own code and directly access MODIS data using R (www.r-project.org/) or Python (www.python.org/) programing environments. Freely available programing resources for handling MODIS data can be found in the GitHub platform for developers (https://github.com/). Both those freely available programing tools, i.e., R and Python as well as other commercial platforms like Matlab (www.mathworks.com/), support functions and provide packages to access, subset both temporally and spatially, reproject, resize and mosaic, scale, and convert to time series of MODIS products. A convenient and user-friendly R package to direct access and preprocess Collection 5 and 6 of MODIS land products is the MODIStsp package (Busetto and Ranghetti 2019). After preprocessing, a tedious task has to take place, and this is data filtering. Filtering is essential for ensuring that only data of good quality are kept for further use in computations and consequently that any assessment based on those data is reliable and scientifically sound.

Data Handling and Filtering

After downloading the requested LST data either in Hierarchical Data Format (HDF) or Geotif format, the working directory will look similar to Table 2.1, where a screenshot of the 11 Geotif files of the MOD11A2 (8-Day Land Surface Temperature/3-Band Emissivity from Terra at 1 km spatial resolution) for day 49 of 2000 is provided. 8-Day LST is the averaged LSTs of the MxD11A1 product over eight days, where the date in the granule is the first day of data in the eight-day composites.

TABLE 2.1   List of layers for MOD11A2 corresponding to 49th day of the year 2000*

MOD11A2.A2000049.h19v04.006.2015058135052_Clear_sky_days
MOD11A2.A2000049.h19v04.006.2015058135052_Clear_sky_nights
MOD11A2.A2000049.h19v04.006.2015058135052_Day_view_angle
MOD11A2.A2000049.h19v04.006.2015058135052_Day_view_time
MOD11A2.A2000049.h19v04.006.2015058135052_Emis_31
MOD11A2.A2000049.h19v04.006.2015058135052_Emis_32
MOD11A2.A2000049.h19v04.006.2015058135052_LST_Day_1km
MOD11A2.A2000049.h19v04.006.2015058135052_LST_Night_1km
MOD11A2.A2000049.h19v04.006.2015058135052_Night_view_angl
MOD11A2.A2000049.h19v04.006.2015058135052_Night_view_time
MOD11A2.A2000049.h19v04.006.2015058135052_QC_Day
MOD11A2.A2000049.h19v04.006.2015058135052_QC_Night

* Data were subset and downloaded using the R package MODIStsp (Busetto and Ranghetti 2019).

Notice that the second part of the filename i.e. A2000049, contains the Day Of Year (DOY) of the eight-day compositing period, and the third part, i.e. h19v04, locates horizontally (from 0 to 35) and vertically (from 0 to 17) the corresponding MODIS tile. The collection number is stored in the next three digits, whereas the last part of the filename contains the production date and time. In both Aqua and Terra 8-Day LST products (MxD11A2) the daytime and nighttime products are combined into a single product which contains different data sets for LST, Quality Control (QC), View angle, and View time for day and night respectively (Hulley et al. 2016). Thus, the entire data set consisted of both daytime and nighttime files and covered the period from the 49th day of 2000 to the 49th day of 2019 for an area in northeast Greece, i.e. 874 dates. The total number of files for the whole period for both daytime and nighttime LST is 10,488, and their size is approximately 8.5 MB for a geographic area of 7 km × 7 km (i.e. 49 km²). Data can be downloaded in Hierarchical Data Format (HDF) as well, with each file containing

TABLE 2.2   Extracting daytime LST and QC layers

```
library(raster)
rlist=list.files(getwd(), pattern=glob2rx("*LST_Day*.tif"), full.names=FALSE)
# reads only LST files from the list of files in the working directory
rlist_QC=list.files(getwd(), pattern=glob2rx("*QC_Day*.tif"), full.
names=FALSE)
# reads only QC files from the list of files in the working directory
for(i in rlist) { assign(unlist(strsplit(i, ".tif")), raster(i)) } # creates raster for each
LST file
for(i in rlist_QC) { assign(unlist(strsplit(i, ".tif")), raster(i)) } # creates raster for
each QC file
a<-stack(rlist) # creates a raster stack for LST layers
a<-a*0.02 # scale factor for Kelvin
b<-stack(rlist_QC) # creates a raster stack for QC layers
# We no longer need raster layers in the memory, so we may now remove them:
rm(list = ls()[grepl("MOD", ls())])
```

11 layers corresponding to the 11 Geotifs for each daytime and nighttime date.

In Table 2.2 a simple R-code is provided for extracting only daytime LST and QC layers from the whole list of files. Handling of raster files is performed through raster R package (Hijmans 2017). All text after the # symbol is comment and is not executed.

Now we have two stacks stored in memory, i.e. a and b, each one holding 874 layers. Each layer corresponds to a specific date.

Before users proceed to the filtering process, it is strongly recommended that they go through the description of the product under study and the associated Scientific Data Sets (SDS) it contains in the provided user's guide. In this case we should refer to the Collection 6 MODIS Land Surface Temperature Products Users' Guide (Zhengming Wan 2013) to acquire information of the data type, units, valid range, fill value, and scale factor for each SDS. In that report, Table 9 provides the data type for the QC layers which for this specific product is 8 bits unsigned integer. Further, we need information on how to decode those 8-bit flags of the quality layer. We may find the decoding information in Table 13 of Zhengming Wan (2013). Now it is easy to write a few lines of R-code to reject those pixels which are not of good quality from each layer of the LST stack. Since the QC layers are in 8-bit unsigned integer type, a total of 256 possible combinations can be defined through the following R lines in Table 2.3 (also available from the NASA MODIS team in GitHub (https://github.com/ornldaac/modis_webservice_qc_filter_R/):

TABLE 2.3 Construction of the QC table

```
# Here an empty table with 256 rows and 13 columns is created to store quality flags
QC_Data <- data.frame(
Integer_Value = 0:255, Bit7 = NA, Bit6 = NA, Bit5 = NA, Bit4 = NA, Bit3 = NA,
Bit2 = NA, Bit1 = NA, Bit0 = NA, Mandatory_QA = NA, Data_Quality = NA,
Emiss_Err = NA, LST_Err = NA)
# Loop through each 8-bit integer and assign 0s and 1s to the QC bit columns
for (i in QC_Data$Integer_Value) {
AsInt <- as.integer(intToBits(i)[1:8])
QC Data[i+1,2:9] <- AsInt[8:1] # Flip to big endian}
# Describe the QC bits based on the criteria outlined in the MODIS product
table
QC_Data$Mandatory_QA[QC_Data$Bit1==0 & QC_Data$Bit0==0] <- 'LST
GOOD'
QC_Data$Mandatory_QA[QC_Data$Bit1==0 & QC_Data$Bit0==1] <- 'LST
Produced,Other Quality'
QC_Data$Mandatory_QA[QC_Data$Bit1==1 & QC_Data$Bit0==0] <- 'No
Pixel,clouds'
QC_Data$Mandatory_QA[QC_Data$Bit1==1 & QC_Data$Bit0==1] <- 'No
Pixel, Other QA'
QC_Data$Data_Quality[QC_Data$Bit3==0 & QC_Data$Bit2==0] <- 'Good
Data'
QC_Data$Data_Quality[QC_Data$Bit3==0 & QC_Data$Bit2==1] <- 'Other
Quality'
QC_Data$Data_Quality[QC_Data$Bit3==1 & QC_Data$Bit2==0] <- 'TBD'
QC_Data$Data_Quality[QC_Data$Bit3==1 & QC_Data$Bit2==1] <- 'TBD'
QC_Data$Emiss_Err[QC_Data$Bit5==0 & QC_Data$Bit4==0] <- 'Emiss Error
<= .01'
```

(Continued)

TABLE 2.3   (Continued) Construction of the QC table

QC_Data$Emiss_Err[QC_Data$Bit5==0 & QC_Data$Bit4==1] <- 'Emiss Err <=.02'

QC_Data$Emiss_Err[QC_Data$Bit5==1 & QC_Data$Bit4==0] <- 'Emiss Err <=.04'

QC_Data$Emiss_Err[QC_Data$Bit5==1 & QC_Data$Bit4==1] <- 'Emiss Err > .04'

QC_Data$LST_Err[QC_Data$Bit7==0 & QC_Data$Bit6==0] <- 'LST Err <=1K'

QC_Data$LST_Err[QC_Data$Bit7==0 & QC_Data$Bit6==1] <- 'LST Err <=3K'

QC_Data$LST_Err[QC_Data$Bit7==1 & QC_Data$Bit6==0] <- 'LST Err <=2K'

QC_Data$LST_Err[QC_Data$Bit7==1 & QC_Data$Bit6==1] <- 'LST Err > 3K'

Now it's time to decide which filtering criteria best fit our study. In this case we will filter pixels that were not produced due to cloud cover (Integer_Value = 2 or 3) and pixels of "other quality" with LST error > 2K (Bit0 = 1 and Bit1 = 0 and Bit6 not equal to 0). It should be mentioned that each research work has its own needs in terms of pixel filtering, and the scientists should define the filtering criteria based on different study type. After the filtering criteria are defined, the QC table is subset to include only rows that correspond to criteria for pixels we wish to exclude from further computations (Table 2.4). The result of the filtering procedure can be viewed in Figure 2.1.

Using MODIS LST Data for Environmental Assessments

After filtering, the time series of images is ready for environmental assessments, and numerous research works have been published either at the global scale (Jin and Dickinson 2010; Mao et al. 2017), at the country scale (Eleftheriou et al. 2018; Zhou et al. 2016), or for regional (Urqueta et al. 2018; Hutengs and Vohland 2016) and local impact studies (Huidong Li et al. 2019). An example of a country wide study in Greece aiming to determine the

TABLE 2.4   Setting the filtering criteria

```
# Filtering procedure
QC_Data <- QC_Data[QC_Data$Bit0 == 1 & QC_Data$Bit1 == 0 & QC_
Data$Bit6 != 0 | QC_Data$Integer_Value %in% c(2, 3),]
rownames(QC_Data) <- NULL
filter <- QC_Data$Integer_Value
b <- b %in% filter
a_filtered<-mask(a,b, maskvalue = 1)
```

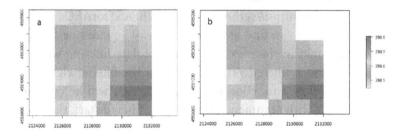

FIGURE 2.1   Daytime LST (in degrees Kelvin) of the MOD11A2 product (eight-day composites) for day 57 of 2000 in a 49 km² area in northeast Greece, (a) before filtering, (b) after filtering. Coordinates are in MODIS Sinusoidal Projection.

annual and seasonal daytime and nighttime trends of MODIS LST and assess potential climate change implications is published in Eleftheriou et al. (2018). Most times, country wide applications need data from more than one MODIS tile. In the case of Greece, three MODIS tiles, i.e. h19v04, h19v05, and h20v05, were downloaded, subset, and filtered for constructing the time series of daytime and nighttime LST from 2000 to 2017, following the procedure described in the previous section. Since Collection 6 was not available at the time of this research, the Collection 5 of MODIS Level 3 LST and Emissivity 8-Day (MOD11A2) product, which averages the daily 1 km LST product of clear sky LSTs during an eight-day period, was used. The objective of the study was to determine the long term trend of daytime and nighttime LST during the 17-year period. In this case the authors have selected a least-squares linear regression model to detect trend (Maselli 2004; Piao et al. 2011). This is a straightforward way to find simple trends over the whole time series. A robust method for performing time series analysis and decomposing it into trend, season, and remainder components, and analyze seasonality and changes within the time series, is the Breaks For Additive Season and Trend (BFAST) method (Verbesselt et al. 2010; Verbesselt, Hyndman, and Culvenor 2010; Verbesselt and Herold 2012), and a comprehensive tutorial, including R-codes, on how to apply it can be found at https://verbe039.github.io/BFASTforAEO/. In the following, users can find an R-code for least square line fitting. Remember that in the previous section of data filtering we had computed the a_filtered raster stack with the filtered eight-day LST layers. Now a function for trend detection will be created, named fun2, with argument x being the a_filtered stack that will be passed later to fun2 to perform linear fitting on a pixel basis through the calc() function (Table 2.5).

In all cases, the user should always keep in mind issues related to the computational restrictions due to the large volume of such a data set, especially when extended geographical areas are examined and large raster stacks are involved. Some tips to

TABLE 2.5   Trend detection function

```
time <- 1:nlayers(a_filtered) # determines time steps
# Function to detect time trends
fun2 <- function(x) {
if (is.na(x[1])){ # detects if all pixels of the time series are NA, e.g. in sea areas
return(cbind(NA,NA,NA)) # in case all pixels are NA then the result will be NA }
 m = lm(x~time) # constructs the linear model
s <- summary(m) # a variable that holds the statistics of the regression
sl <- s$coefficients[2,1] # computes slope of the linear model
r2 <- s$r.squared # computes the r-squared of the regression
# the next line detects statistical significance of the model (p value)
pf<- pf(s$fstatistic[1], s$fstatistic[2], s$fstatistic[3],lower.tail = FALSE)
cbind(r2, pf, sl) # holds together
} # end of fun2
```

reduce computational effort and therefore time of computations are related to use of parallel processing. An example of an R-code for multicore processing using the snow R package (Tierney et al. 2018) is given in Table 2.6.

Alternatively, for small areas the user may go through the regression computations (Table 2.7) directly without parallel processing.

Results of the computation of time trends in the 7 km x 7 km area of the previous section in northeast Greece can be seen in Figure 2.2. Users should note that only a small portion of pixels provide statistically significant trends. This is due mostly to pixels removed from the time series because of cloud cover conditions

TABLE 2.6   Multicore processing R-code

```
beginCluster(4) # a cluster of four processor is initiated
# the next line uses clusterR function to apply fun2 to each pixel
z1 <- clusterR(a_filtered, calc, args=list(fun=fun2), export='time')
endCluster()
```

TABLE 2.7   Regression computations

```
# Application of fun2 using the calc function of the raster package. Calc applies a function on
# a pixel basis from a raster object of a multi-layer raster e.g. stack. r3 is a raster stack with
# three layers. The first one holds the r-squared of the regression line for each pixel,
# the second the p-values of statistical significance, and the third layer the slope of
# the regression line.
r3 <- calc(a_filtered, fun2)
r3[r3[[2]] > 0.05] <- NA # deletes pixel values with p-value > 0.05
plot(r3) # plots the three layers of r3 stack. See Figure 2.2.
writeRaster(r3[[3]], "sl", format = "GTiff") # writes the raster of slope values in Geotif
me<-cellStats(r3[[3]],mean) # computes the mean value of statistically significant trends in
# the study area
```

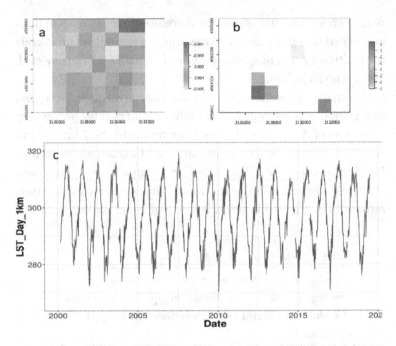

FIGURE 2.2   Daytime LST slope of the trend line (in degrees Kelvin/8 days) (a) before and (b) after removing pixels with statistically insignificant trends (p-values > 0.05), and (c) graph of the subset mean for the whole period. Coordinates are in MODIS Sinusoidal Projection, LST is in degrees Kelvin.

and is the main disadvantage of remotely sensed products which are computed using optical and near-infrared spectral windows.

The same methodology and codes were applied for the three MODIS tiles covering Greece on both annual and seasonal basis. Results published in Eleftheriou et al. (2018), showed a general daytime LST decrease all over the country annually and during all seasons, except winter. Nighttime LST, which corresponds to minimum LST, demonstrated increasing trend on annual and seasonal basis, indicating that both minimum annual and seasonal LST decrease. This pattern of increasing nighttime and decreasing daytime LST corresponds to decrease of diurnal range of

LST which can be regarded as an indicator of climate change. In general, decrease of either air temperature or LST diurnal range indicates possible climate change effects whereas surface conditions like intense urbanization can also be considered as possible driving factors (Qu, Wan, and Hao 2014; Sun, Pinker, and Kafatos 2006; Zhou et al. 2016).

Besides the important advantages of satellite observations, users should always keep in mind that there are some limitations concerning specific remotely sensed products. As such, cloud cover, atmospheric conditions, topography, and spatial resolution are the most common limitations recognized by scientists (Jin and Dickinson 2010). Specifically, for climate assessments using remotely sensed LST products, scientists should be aware of possible biases in related computations, the most obvious being the availability of observations only during clear sky conditions. Furthermore, when mean daily temperature is the target variable, users should consider that estimation of mean daily temperature as an average of two instant measurements which is the case of MODIS and other LST products introduces a certain bias (Dall'Amico and Hornsteiner 2006). Nevertheless, recent studies estimated mean surface temperature at the global scale using MODIS data. Mao et al. (2017), using a 12-year period, from 2001 to 2012, of global MODIS LST data, concluded that there cannot be a definite answer for global warming because although most regions in the Southern Hemisphere seem to go warmer, there are many regions which seem to be cooling, like central and eastern areas of the Pacific Ocean, northern regions of the Atlantic Ocean, northern China, Mongolia, southern regions of Russia, western regions of Canada and America, the eastern and northern regions of Australia, and the southern part of Africa. Furthermore, it should be highlighted that the time series of available remotely sensed data cover at best a 30-year period, and even in those cases, assessments for climate change implications should be made cautiously and should be considered as potentially indicative of climate changes but not as conclusive evidences.

2.3 MODIS NDVI PRODUCT: TRENDS OF NDVI CHANGES IN VARIOUS LAND COVER AREAS

While scientists deal with environmental and climate assessments, they usually need information of vegetation status of study regions. This is because vegetation has a mitigative role on climate change impacts and is recognized as an important CO_2 sink, even greater than oceans (Quéré et al. 2009). In that work, Quéré et al. (2009), indicated that greenhouse gas (GHG) emissions from certain changes of land uses such as deforestation, urbanization, and intensive agricultural activities, among others, are the second largest CO_2 sources. Nevertheless, there are other land use changes that partially compensate for CO_2 emissions due to land use changes. Such land use changes are afforestation, growing of urban vegetation, and soil carbon conservative cultivation practices. Ma et al. (2013) indicated the important role of the correct representation of landscape phenology on the modeling results of climate models whereas Li et al. (2019) came to the same conclusion where urban climate modeling is concerned.

Terrestrial ecosystems experience vegetation changes which are evidenced mainly as upward migration of plant species, changing of growing period, productivity, and phenology (Parmesan and Yohe 2003; Cong et al. 2013). They are known to take place due to interaction of global scale drivers such as rise of global temperature and atmospheric CO_2 concentration with local scale drivers such as local land use changes or catastrophic events such as wild fires or floods. It is thus recognized that monitoring vegetation quality is of primary importance for local and global climate assessment studies and for designing adaptation policies (Alexandra Gemitzi, Banti, and Lakshmi 2019).

Various studies of terrestrial vegetation have been conducted for different areas of the globe during the last 20 years. Most of those indicate a greening vegetation pattern in China (Xu et al. 2014), Central Asia (Yin et al. 2016), and the Himalayas (Mishra and Chaudhuri 2015; Haidong Li et al. 2016; Mishra and Mainali 2017). Mixed trends of terrestrial vegetation either increasing or decreasing have been reported in the African savanna (Mishra et al. 2015), with greening trends attributed to increased moisture availability and negative vegetation trends associated with mega fire events. Also in boreal Eurasia, Piao et al. (2011) reported mixed patterns of increasing trends before 2000 and decreasing vegetation up to 2006, mostly during summers linked to decreased precipitation during that season. Similar mixed vegetation patterns were highlighted by Liu et al. (2016) in the Tianshan Mountains in China, attributed to variability of hydroclimatic parameters such as temperature, precipitation, and soil moisture.

It seems therefore of unique importance to measure and quantify the role of terrestrial vegetation in controlling environmental and mostly climate processes at the global scale. Consequently, scientists need to monitor the distribution and spatio-temporal variations of vegetation. From this point of view remote sensing offers indisputable advantages for large scale monitoring, providing robust indices of vegetation activity, known as Vegetation Indices (VI) (Didan, Munoz, and Huete 2015). Among the numerous remotely sensed VI products MODIS VI products are considered of high quality as they been successfully evaluated in most land cover types (Huete et al. 2002), and they are also used as a state of the art data set to assess the quality and performance of VI from other sensors (Kern, Marjanović, and Barcza 2016). Remotely sensed VI products combine two or more atmospherically corrected bands in the Red, Blue, and Near Infrared (NIR) wavelengths, thus enhancing the reflected signal from vegetation. Commonly, two VI products are released from various remote sensing missions, the Normalized Difference Vegetation Index (NDVI) and the Enhanced Vegetation Index (EVI). NDVI is known to saturate in high biomass areas where EVI offers improved performance (Huete et al. 2002). The processing algorithms, theoretical background, quality assurance issues, and scientific data sets are described in detail in Didan, Munoz, and Huete (2015). MODIS VI products are released only for land areas at a spatial resolution of 250 m, 500 m, 1 km, and 0.05 degrees. Eight gridded MODIS products are released at a 16-day temporal resolution, i.e. Vegetation Indices 16-Day L3 Global 250m (MOD13Q1/MYD13Q1), Vegetation Indices 16-Day L3 Global 500m (MOD13A1/MYD13A1), Vegetation Indices 16-Day L3 Global 1km (MOD13A2/MYD13A2), and Vegetation Indices 16-Day L3 Global 0.05Deg CMG (MOD13C1/MYD13C1), and four more at a monthly time scale, i.e. Vegetation Indices Monthly L3 Global 1km (MOD13A3/MYD13A3), Vegetation Indices Monthly L3 Global 0.05Deg CMG (MOD13C2/MYD13C2). Since optical and NIR wavelengths are involved, known issues are related to cloud or snow cover of the target areas or aerosol contamination in the atmosphere.

In the following example an application of vegetation trend assessment for an area in northeast Greece will be presented using the Collection 6 MOD13Q1 product (Didan, Munoz, and Huete 2015), which comprises 16-day NDVI/EVI data at 250 m spatial resolution.

For a typical Mediterranean country like Greece, the NDVI is suitable for assessments concerning vegetation patterns; therefore, the computations in the following sections are for NDVI only, although similar codes can be applied for EVI as well.

Data Handling and Filtering

Concerning data access for MODIS VI products, the information provided in section 2.2 for MODIS LST products is also applicable for the case of VI. In this case as well data were subset and downloaded using the R package MODIStsp (Busetto and Ranghetti 2019). The Collection 6 of MOD13Q1 product comprises 12 scientific data sets: 250m 16-days NDVI, 250m 16-days EVI, 250m 16-days VI Quality, 250m 16-days Red reflectance (Band 1), 250m 16-days NIR reflectance (Band 2), 250m 16-days Blue reflectance (Band 3), 250m 16-days MIR reflectance (Band 7), 250m 16-days View Zenith Angle, 250m 16-days Sun Zenith Angle, 250m 16-days Relative Azimuth Angle, 250m 16-days Composite Day Of the Year, and 250m 16-days Pixel Reliability. For a time series covering the period from the 49th day of year 2000 to the 49th day of year 2019, the total number of Geotif files is 5,256 and the total size is approximately 15.5 MB. A list of data type, units, valid range, and scale factor can be found in Table 1 of MODIS Vegetation Index User's Guide (MOD13 Series) (Didan, Munoz, and Huete 2015). Quality assurance related science data sets are 250m 16-days VI Quality and the 250m 16-days Pixel Reliability. The latter contains simplified overall pixel quality. The VI Quality layer provides information for both NDVI and EVI since evaluation of earlier versions of those products revealed insignificant quality differences among NDVI and EVI. This layer is encoded in 16-bit unsigned integer format, and decoding information in provided in Table 5 of Didan, Munoz, and Huete (2015). Users should keep in mind that in contrast to MODIS LST product examined in section 2.2 which was encoded in 8-bit unsigned integer, the VI quality layer provides 65,535 (2^{16}) different quality possibilities, and their decoding should follow a somehow different strategy than that of Tables 2.3 and 2.4. This time we will not construct the whole QC table. Instead, for each time step we will produce a different QC table, with one row for each unique integer value in each QC layer. The code is provided in Table 2.8, together with lines for rasterizing, stacking, and scaling

TABLE 2.8   Rasterizing, stacking, scaling, and filtering of MODIS NDVI product

```
library(raster)
# read all NDVI layers and VI Quality layers in the working directory
rlist=list.files(getwd(), pattern=glob2rx("*NDVI*.tif"), full.names=FALSE)
rlist_QC=list.files(getwd(), pattern=glob2rx("*VI_Quality*.tif"), full.names=FALSE)
for(i in rlist) { assign(unlist(strsplit(i, ".tif")), raster(i)) } # create rasters for NDVI layers
for(i in rlist_QC) { assign(unlist(strsplit(i, ".tif")), raster(i)) }#create rasters for Quality layers
a<-stack(rlist) # creates a raster stack for NDVI layers
a<-a*0.0001 # scale factor for NDVI            ~
b<-stack(rlist_QC) #creates a raster stack for Quality layers
# Now we may remove all rasters from memory
```

(*Continued*)

TABLE 2.8   (Continued) Rasterizing, stacking, scaling, and filtering of MODIS NDVI product

---

rm(list = ls()[grepl("MOD", ls())])# removes all objects with MOD in their name

Filtering using the 16-bit unsigned integer coded quality layer for each date

for (i in 1:length(rlist)) { # loop through each date of NDVI layer
qcvals<-unique(b[[i]]) # read all unique values in each quality layer
QC_Data <- data.frame(

# A column with one row for each unique integer value in the QC layer time series
Integer_Value = qcvals,
# An empty column for each bit
Bit15 = NA, Bit14 = NA, Bit13 = NA, Bit12 = NA, Bit11 = NA, Bit10 = NA, Bit9 = NA, Bit8 = NA,
Bit7 = NA, Bit6 = NA, Bit5 = NA, Bit4 = NA, Bit3 = NA, Bit2 = NA, Bit1 = NA, Bit0 = NA,

# An empty column for each QC group as defined in Table 5 of Didan, Munoz, and Huete (2015)
MODLAND_QA = NA, VI_usefulness = NA, Aerosol_quantity = NA,
Adjacent_cloud_detected = NA, Atmosphere_BRDF_correction_performed = NA,
Mixed_clouds = NA, Land_Water_flag = NA, Possible_snow_ice = NA, Possible_shadow =
NA)

for(j in seq(1,length(QC_Data$Integer_Value),1)){
 AsInt <- as.integer(intToBits(QC_Data[j,])[1:16])
 QC_Data[j,2:17] <- AsInt[16:1] # Flip to big endian }

Describe the QC bits based on the criteria outlined in the MODIS product table
QC_Data$MODLAND_QA[QC_Data$Bit1==0 & QC_Data$Bit0==0] <- 'VI produced, good quality'
QC_Data$MODLAND_QA[QC_Data$Bit1==0 & QC_Data$Bit0==1] <- 'VI produced, but check other QA'
QC_Data$MODLAND_QA[QC_Data$Bit1==1 & QC_Data$Bit0==0] <- 'Pixel produced, but most probably cloudy'
QC_Data$MODLAND_QA[QC_Data$Bit1==1 & QC_Data$Bit0==1] <- 'Pixel not produced due to other reasons than clouds'

QC_Data$VI_usefulness[QC_Data$Bit5==0 & QC_Data$Bit4==0 & QC_Data$Bit3==0 & QC_Data$Bit2==0] <- 'Quality Lv 1'
QC_Data$VI_usefulness[QC_Data$Bit5==0 & QC_Data$Bit4==0 & QC_Data$Bit3==0 & QC_Data$Bit2==1] <- 'Quality Lv 2'
QC_Data$VI_usefulness[QC_Data$Bit5==0 & QC_Data$Bit4==0 & QC_Data$Bit3==1 & QC_Data$Bit2==0] <- 'Quality Lv 3'

(*Continued*)

TABLE 2.8 (Continued) Rasterizing, stacking, scaling, and filtering of MODIS NDVI product

QC_Data$VI_usefulness[QC_Data$Bit5==0 & QC_Data$Bit4==0 & QC_Data$Bit3==1 & QC_Data$Bit2==1] <- 'Quality Lv 4'

QC_Data$VI_usefulness[QC_Data$Bit5==0 & QC_Data$Bit4==1 & QC_Data$Bit3==0 & QC_Data$Bit2==0] <- 'Quality Lv 5'

QC_Data$VI_usefulness[QC_Data$Bit5==0 & QC_Data$Bit4==1 & QC_Data$Bit3==0 & QC_Data$Bit2==1] <- 'Quality Lv 6'

QC_Data$VI_usefulness[QC_Data$Bit5==0 & QC_Data$Bit4==1 & QC_Data$Bit3==1 & QC_Data$Bit2==0] <- 'Quality Lv 7'

QC_Data$VI_usefulness[QC_Data$Bit5==0 & QC_Data$Bit4==1 & QC_Data$Bit3==1 & QC_Data$Bit2==1] <- 'Quality Lv 8'

QC_Data$VI_usefulness[QC_Data$Bit5==1 & QC_Data$Bit4==0 & QC_Data$Bit3==0 & QC_Data$Bit2==0] <- 'Quality Lv 9'

QC_Data$VI_usefulness[QC_Data$Bit5==1 & QC_Data$Bit4==0 & QC_Data$Bit3==0 & QC_Data$Bit2==1] <- 'Quality Lv 10'

QC_Data$VI_usefulness[QC_Data$Bit5==1 & QC_Data$Bit4==0 & QC_Data$Bit3==1 & QC_Data$Bit2==0] <- 'Quality Lv 11'

QC_Data$VI_usefulness[QC_Data$Bit5==1 & QC_Data$Bit4==0 & QC_Data$Bit3==1 & QC_Data$Bit2==1] <- 'Quality Lv 12'

QC_Data$VI_usefulness[QC_Data$Bit5==1 & QC_Data$Bit4==1 & QC_Data$Bit3==0 & QC_Data$Bit2==0] <- 'Quality Lv 13'

QC_Data$VI_usefulness[QC_Data$Bit5==1 & QC_Data$Bit4==1 & QC_Data$Bit3==0 & QC_Data$Bit2==1] <- 'Quality so low that it is not useful'

QC_Data$VI_usefulness[QC_Data$Bit5==1 & QC_Data$Bit4==1 & QC_Data$Bit3==1 & QC_Data$Bit2==0] <- 'L1B data faulty'

QC_Data$VI_usefulness[QC_Data$Bit5==1 & QC_Data$Bit4==1 & QC_Data$Bit3==1 &

QC_Data$Bit2==1] <- 'Not useful for any other reason/not processed'

QC_Data$Aerosol_quantity[QC_Data$Bit7==0 & QC_Data$Bit6==0] <- 'Climatology'

QC_Data$Aerosol_quantity[QC_Data$Bit7==0 & QC_Data$Bit6==1] <- 'Low'

QC_Data$Aerosol_quantity[QC_Data$Bit7==1 & QC_Data$Bit6==0] <- 'Average'

QC_Data$Aerosol_quantity[QC_Data$Bit7==1 & QC_Data$Bit6==1] <- 'High'

QC_Data$Adjacent_cloud_detected[QC_Data$Bit8==0] <- 'No'

QC_Data$Adjacent_cloud_detected[QC_Data$Bit8==1] <- 'Yes'

QC_Data$Atmosphere_BRDF_correction_performed[QC_Data$Bit9==0] <- 'No'

(Continued)

TABLE 2.8   (Continued) Rasterizing, stacking, scaling, and filtering of MODIS NDVI product

---

```
QC_Data$Atmosphere_BRDF_correction_performed[QC_Data$Bit9==1] <- 'Yes'
QC_Data$Mixed_clouds[QC_Data$Bit10==0] <- 'No'
QC_Data$Mixed_clouds[QC_Data$Bit10==1] <- 'Yes'
QC_Data$Land_Water_flag[QC_Data$Bit13==0 & QC_Data$Bit12==0 &
QC_Data$Bit11==0] <- 'Land and desert'
QC_Data$Land_Water_flag[QC_Data$Bit13==0 & QC_Data$Bit12==0 &
QC_Data$Bit11==1] <- 'land no desert'
QC_Data$Land_Water_flag[QC_Data$Bit13==0 & QC_Data$Bit12==1 &
QC_Data$Bit11==0] <- 'inland water'
QC_Data$Land_Water_flag[QC_Data$Bit13==0 & QC_Data$Bit12==1 &
QC_Data$Bit11==1] <- 'sea water'
QC_Data$Land_Water_flag[QC_Data$Bit13==1 & QC_Data$Bit12==0 &
QC_Data$Bit11==1] <- 'coastal'
QC_Data$Possible_snow_ice[QC_Data$Bit14==0] <- 'No'
QC_Data$Possible_snow_ice[QC_Data$Bit14==1] <- 'Yes'
QC_Data$Possible_shadow[QC_Data$Bit15==0] <- 'No'
QC_Data$Possible_shadow[QC_Data$Bit15==1] <- 'Yes'
# Define which pixels to exclude
QC_Data <- QC_Data[QC_Data$MODLAND_QA %in% MODLAND_QA |
QC_Data$VI_usefulness %in% VI_usefulness |
QC_Data$Aerosol_quantity==Aerosol_quantity |
QC_Data$Adjacent_cloud_detected==Adjacent_cloud_detected |
QC_Data$Possible_shadow==Possible_shadow,]
rownames(QC_Data) <- NULL
filter <- QC_Data$Integer_Value # We will remove these values
b[[i]] <- b[[i]] %in% filter # creating the mask
a_filtered[[i]]<-mask(a[[i]],b[[i]], maskvalue=1)}
```

Geotif files related to NDVI and VI Quality in the list of MODIS layers. EVI layers can be extracted as well, changing only the rlist in Table 2.8. Scaling factor for EVI and filtering procedure are exactly the same for EVI layers. In each different study, however, we have to decide on the filtering criteria, i.e. what pixels to exclude from the analysis using the information provided in the group columns in

Table 5 of Didan, Munoz, and Huete (2015). Only the rows for integers that represent pixels of unacceptable quality are kept. For this specific example, the filtering criteria are set to: MODLAND_QA equal to 'Pixel produced, but most probably cloudy', or 'Pixel not produced due to other reasons than clouds', VI_usefulness equal to "Quality Lv 4" or "Quality Lv 5" or "Quality Lv 6" or "Quality Lv 7" or "Quality Lv 8" or "Quality Lv 9" or "Quality Lv 10" or "Quality Lv 11" or "Quality Lv 12" or "Quality Lv 13" or "Quality so low that it is not useful" or "L1B data faulty" or "Not useful for any other reason/not processed", Aerosol_quantity equal to "High", Adjacent_cloud_detected equal to "Yes", Possible_shadow equal to "Yes".

The same methodology and codes were applied for the three MODIS tiles covering Greece on both annual and seasonal basis from 2000 to 2017, and the analysis was also conducted on the different land use type of the country. The limitations of the methodology are related to the ones defined in section 2.2 regarding the inability of optical/infrared radiation to penetrate clouds. Results were published in Banti, Kiachidis, and Gemitzi (2019) and Gemitzi, Banti, and Lakshmi (2019). Findings of those works document an increasing NDVI trend both annually and seasonally, with a mean annual rate of 1.25×10^{-3} year^{-1}. Compared to the global mean NDVI trend of 0.46×10^{-3} year^{-1} from 1982 to 2012 estimated in Y. Liu et al. (2015) it seems that Greece demonstrates a higher increasing vegetation productivity. Sporadic negative trends were found in areas that had experienced fires and in some areas of intense touristic development. Among the different land use types examined in Greece, the highest rates of increase of vegetation productivity are found in forest and semi-natural areas and in agricultural areas. Only burned areas demonstrated decreasing vegetation trends whereas an interesting finding is the increasing trend found in artificial areas. Analogous findings of greening patterns in urban sites were observed in other areas of the world, e.g. in northern West Siberia (Esau et al. 2016) and in Israel (Levin 2016), and in both cases they were related to targeted efforts towards environmental-friendly residential areas. In Greece, however, we should also take into account the possible positive effects of economic crisis (since 2010) on vegetation, as there was an abrupt decline in all production activities of the country. Other possible driving factors of the observed vegetation trend patterns in Greece were also examined, and a positive correlation of NDVI trends with the increasing nighttime LST trends (described in section 2.2) was revealed in all seasons and in all land use types except burned areas.

2.4 MODIS EVAPOTRANSPIRATION PRODUCT: GROUNDWATER RECHARGE ESTIMATION USING HYDROLOGICAL MODELING AND REMOTELY SENSED DATA

Among the many useful scientific products from MODIS, evapotranspiration (ET) products are perhaps of particular interest to hydrologists as this parameter is difficult to quantify, especially its transpiration fraction. In Greece, which is a typical Mediterranean country, almost 50% of precipitation is transferred back to the atmosphere in the form of evapotranspiration (European Academies Science Advisory Council [EASAC] 2010), and besides its significant contribution in the water budget, it is also very useful in indirectly quantifying groundwater recharge and estimating the renewable groundwater resources. Hydrological models offer an alternative to estimate the various components of the water budget, including groundwater recharge and evapotranspiration, but the accuracy of their output is strongly dependent on the availability of measurements of various water components for calibrating and verifying the model. There are, however, many basins where hydrologic data availability is a problem. Even in cases where there is a monitoring network, its maintenance and proper functioning is costly and demanding, and frequently such networks cannot provide continuous hydrologic observations to support the development of a reliable hydrological model. Application of remotely sensed monitoring data for modeling and predicting the hydrological status of ungauged basins can contribute significantly (Lakshmi 2013; Mohanty, Lakshmi, and Montzka 2013; Lakshmi 2004).

The scientific basis for MOD16 products is found in the Penman–Monteith ET formula (Monteith J L 1965) using auxiliary information from the MODIS land cover, albedo, leaf area index (LAI), and EVI along with daily meteorological reanalysis data. Mu et al. (2007) developed an algorithm that calculates ET using the Penman – Monteith ET formula (Monteith 1964), and the initially developed methodology was further improved to account for nighttime ET components (Mu, Zhao, and Running 2011). MODIS evapotranspiration products are global gridded Level-4 data sets at 500 m spatial resolution (https://modis.gsfc.nasa.gov/ data/dataprod/mod16.php). Two eight-day composites (MOD16A2/ MYD16A2) are provided whereas the two composites at the yearly timescale (MOD16A3/MYD16A3) are currently unavailable due to unexpected errors in the input data. Scientific layers for MOD16A2/ MYD16A2 products correspond to composited Evapotranspiration (ET), Potential ET (PET) (sum of eight days), Latent Heat Flux (LE), and Potential LE (PLE) (average of eight days). Details on the algorithm and scientific data sets of MOD16A2/MYD16A2 products can be found in Running et al. (2019). A QC layer encoded in 8-bit unsigned integer is also provided, but it does not reflect the quality of estimated evapotranspiration. The QC layer contains the quality information of the corresponding input LAI/FPAR (MYD15A2H) granule of the same eight-day composite period. For this reason, users are advised to use cautiously the evapotranspiration products.

Within this section an application of MODIS evapotranspiration product for estimation of groundwater recharge will be presented. The methodology is described in detail in Gemitzi, Ajami, and Richnow (2017) and focuses on the development of a regression equation between monthly groundwater recharge and monthly effective precipitation, i.e. precipitation minus actual evapotranspiration. Since groundwater recharge rate observations are very scarce, scientists usually need to resort to hydrological modeling to acquire modeled groundwater recharge values. Although a requirement of the methodology is the existence of a reliable hydrological model, which is not always the case, it is argued that the methodology and equations can be applied in ungauged areas with similar climate and geological conditions with the reference basin (the basin where the regression model was developed and verified).

Therefore, in the following case, the methodology will be applied in Phyliouris river basin in northeast Greece, an area adjacent to the study area examined in Gemitzi, Ajami, and Richnow (2017). The regression equation for monthly groundwater recharge estimation developed in Gemitzi, Ajami, and Richnow (2017) will be applied for a two-year period, i.e. 2017–2018:

$$MGR = 0.5174 \times (P - ET_{MODIS}) + 0.2145 \qquad (2.1)$$

where MGR is monthly groundwater recharge in mm, P is monthly precipitation in mm, and ET is monthly evapotranspiration in mm derived from MOD16A2 product. P – ET_{MODIS} corresponds to effective precipitation. Monthly precipitation data were acquired from a local meteo station (Komotini city). MOD16A2 data set, filtered and scaled, was accessed through ORNL DAAC, 2018: MODIS and VIIRS Land Products Global Subsetting and Visualization Tool. ORNL DAAC, Oak Ridge, Tennessee, U.S. (https://doi.org/10.3334/ORNLDAAC/1379).

Figure 2.3a shows a location map of the study area along with time series graphs of monthly precipitation, monthly MODIS evapotranspiration, and estimated groundwater recharge for 2017 and 2018 using Equation (2.1). In this case we did not conduct an analysis on a pixel basis, since QC layers for MODIS evapotranspiration do not guarantee reliable pixel filtering based on QC values. Instead a mean of filtered and scaled values of pixels in the study basin was used for computations to acquire an estimate of groundwater recharge at the basin level. Comparing monthly precipitation and MODIS ET on Figure 2.3b, it is obvious that MODIS

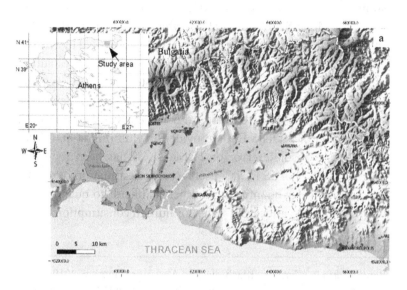

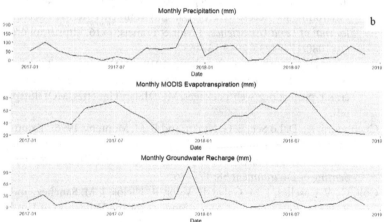

FIGURE 2.3 (a) Location of Phyliouris river basin (light colored dotted line), (b) time series graphs (2017–2018) of monthly precipitation, monthly MODIS evapotranspiration, and estimated monthly groundwater recharge using Eq. (2.1).

ET is considerably higher than precipitation especially during summer months. A possible explanation comes from the algorithm for MODIS ET computation, which uses reanalysis meteorological data with a coarse spatial resolution provided by NASA's Global Modeling and Assimilation Office. The coarse resolution of such reanalysis data, even after being interpolated to reach the spatial resolution of 0.5 km of MOD16 outputs, cannot depict the spatial variability of meteorological parameters at the local scale. A validation of MODIS ET is provided in Running et al. (2019), and a mean absolute error of 24.1%–24.6% has been reported. Even as such, MODIS ET is a useful tool for estimation of renewable groundwater quantity, a parameter difficult to quantify but very useful when allocating water for human consumption.

REFERENCES

Banti, Maria A., Kyriakos Kiachidis, and Alexandra Gemitzi. 2019. "Estimation of Spatio-Temporal Vegetation Trends in Different Land Use Environments Across Greece Use Environments Across Greece." *Journal of Land Use Science.* Taylor & Francis: 1–16. https://doi.org /10.1080/1747423X.2019.1614687.

Busetto, Lorenzo, and Luigi Ranghetti. 2019. "MODIStsp: A Tool for Automatic Preprocessing of MODIS Time Series – v1.3.9." https:// cran.r-project.org/web/packages/MODIStsp/vignettes/MODIStsp. pdf.

Chehbouni, A., D. Lo See, E. G. Njoku, and B. M. Monteny. 1996. "Examination of the Difference Between Radiative and Aerodynamic Surface Temperatures Over Sparsely Vegetated Surfaces." *Remote Sensing of Environment* 58: 177–86.

Coll, C., V. Caselles, J. M. Galve, E. Valor, R. Niclòs, J. M. Sánchez, and R. Rivas. 2005. "Ground Measurements for the Validation of Land Surface Temperatures Derived From AATSR and MODIS Data." *Remote Sensing of Environment* 97: 288–300.

Cong, Nan, Tao Wang, Huijuan Nan, Yuecun Ma, Xuhui Wang, Ranga B. Myneni, and Shilong Piao. 2013. "Changes in Satellite-Derived Spring Vegetation Green-Up Date and Its Linkage to Climate in China From 1982 to 2010: A Multimethod Analysis." *Global Change Biology* 19 (3): 881–891. https://doi.org/10.1111/gcb.12077.

Dall'Amico, M., and M. Hornsteiner. 2006. "A Simple Method for Estimating Daily and Monthly Mean Temperatures From Daily Minima and Maxima." *International Journal of Climatology* 26 (13): 1929–1936. https://doi.org/10.1002/joc.1363.

Didan, Kamel, Armando Barreto Munoz, and Alfredo Huete. 2015. "MODIS Vegetation Index User's Guide (MOD13 Series)." https://vip.arizona.edu/documents/MODIS/MODIS_VI_UsersGuide_June_2015_C6.pdf.

Eleftheriou, Dimitrios, Kyriakos Kiachidis, Georgios Kalmintzis, Argiro Kalea, Christos Bantasis, Paraskevi Koumadoraki, Maria Eleni Spathara, Angeliki Tsolaki, Maria Irini Tzampazidou, and Alexandra Gemitzi. 2018. "Determination of Annual and Seasonal Daytime and Nighttime Trends of MODIS LST Over Greece – Climate Change Implications." *Science of the Total Environment* 616–617. Elsevier B.V.: 937–947. https://doi.org/10.1016/j.scitotenv.2017.10.226.

Esau, Igor, Victoria V. Miles, Richard Davy, Martin W. Miles, and Anna Kurchatova. 2016. "Trends in Normalized Difference Vegetation Index (NDVI) Associated With Urban Development in Northern West Siberia." *Atmospheric Chemistry and Physics* 16 (15): 9563–9577. https://doi.org/10.5194/acp-16-9563-2016.

European Academies Science Advisory Council (EASAC). 2010. "Groundwater in the Southern Member States of the European Union: An Assessment of Current Knowledge and Future Prospects Country Report for Greece Contents Greece Groundwater Report." *EASAC.*

Gallo, Kevin, Robert Hale, Dan Tarpley, and Yunyue Yu. 2011. "Evaluation of the Relationship Between Air and Land Surface Temperature Under Clear- and Cloudy-Sky Conditions." *Journal of Applied Meteorology and Climatology* 50 (3): 767–775. https://doi.org/10.1175/2010JAMC2460.1.

Gemitzi, Alexandra, H. Ajami, and H. H. Richnow. 2017. "Developing Empirical Monthly Groundwater Recharge Equations Based on Modeling and Remote Sensing Data – Modeling Future Groundwater Recharge to Predict Potential Climate Change Impacts." *Journal of Hydrology* 546: 1–13. https://doi.org/10.1016/j.jhydrol.2017.01.005.

Gemitzi, Alexandra, Maria A. Banti, and Venkat Lakshmi. 2019. "Vegetation Greening Trends in Different Land Use Types: Natural Variability Versus Human-Induced Impacts in Greece." *Environmental Earth Sciences* 78 (5). Springer Berlin Heidelberg: 1–10. https://doi.org/10.1007/s12665-019-8180-9.

Hijmans, Robert J. 2017. "Introduction to the 'Raster' Package (Version 2.3–24)." *R-CRAN Project.*

Huete, A., K. Didan, T. Miura, E. P. Rodriguez, X. Gao, and L. G. Ferreira. 2002. "Overview of the Radiometric and Biophysical Performance of the MODIS Vegetation Indices." *Remote Sensing of Environment* 83 (1–2). Elsevier: 195–213. https://doi.org/10.1016/S0034-4257(02)00096-2.

Hulley, G., R. Freepartner, N. Malakar, and S. Sarkar. 2016. "Moderate Resolution Imaging Spectroradiometer (MODIS) Land Surface Temperature and Emissivity Product (MxD21) User Guide." https://modis.gsfc.nasa.gov/data/user_guide/atbd_mod21_userguide.pdf.

Hutengs, Christopher, and Michael Vohland. 2016. "Downscaling Land Surface Temperatures at Regional Scales With Random Forest Regression." *Remote Sensing of Environment* 178. Elsevier Inc.: 127–141. https://doi.org/10.1016/j.rse.2016.03.006.

Jin, Menglin, and Robert E Dickinson. 2010. "Land Surface Skin Temperature Climatology: Benefitting From the Strengths of Satellite Observations." *Environmental Research Letters* 5 (4): 044004. https://doi.org/10.1088/1748-9326/5/4/044004.

Kern, Anikó, Hrvoje Marjanović, and Zoltán Barcza. 2016. "Evaluation of the Quality of NDVI3g Dataset Against Collection 6 MODIS NDVI in Central Europe Between 2000 and 2013." *Remote Sensing* 8 (11). https://doi.org/10.3390/rs8110955.

Lakshmi, Venkat. 2004. "The Role of Satellite Remote Sensing in the Prediction of Ungauged Basins." *Hydrological Processes* 18: 1029–1034. https://doi.org/10.1002/hyp.5520.

Lakshmi, Venkat. 2013. "Remote Sensing of Soil Moisture." *ISRN Soil Science.* Article ID 424178. https://doi.org/10.1155/2013/424178.

Levin, Noam. 2016. "Human Factors Explain the Majority of MODIS-Derived Trends in Vegetation Cover in Israel: A Densely Populated Country in the Eastern Mediterranean." *Regional Environmental Change* 16 (4). Springer Berlin Heidelberg: 1197–1211. https://doi.org/10.1007/s10113-015-0848-4.

Li, Haidong, Jiang Jiang, Bin Chen, Yingkui Li, Yuyue Xu, and Weishou Shen. 2016. "Pattern of NDVI-Based Vegetation Greening Along an Altitudinal Gradient in the Eastern Himalayas and Its Response to Global Warming." *Environmental Monitoring and Assessment* 188 (3): 1–10. https://doi.org/10.1007/s10661-016-5196-4.

Li, Huidong, Yuyu Zhou, Xun Wang, Xu Zhou, Huiwen Zhang, and Sahar Sodoudi. 2019. "Science of the Total Environment Quantifying Urban Heat Island Intensity and Its Physical Mechanism Using

WRF/UCM." *Science of the Total Environment* 650. Elsevier B.V.: 3110–3119. https://doi.org/10.1016/j.scitotenv.2018.10.025.

Liu, Qun, Zhaoping Yang, Fang Han, Zhaoguo Wang, and Cuirong Wang. 2016. "NDVI-Based Vegetation Dynamics and their Response to Recent Climate Change: A Case Study in the Tianshan Mountains, China." *Environmental Earth Sciences* 75 (16). Springer Berlin Heidelberg: 1–15. https://doi.org/10.1007/s12665-016-5987-5.

Liu, Ya, Yan Li, Shuangcheng Li, and Safa Motesharrei. 2015. "Spatial and Temporal Patterns of Global NDVI Trends: Correlations With Climate and Human Factors." *Remote Sensing* 7 (10): 13233–13250. https://doi.org/10.3390/rs71013233.

Lu, Lei, Tingjun Zhang, Tiejun Wang, and Xiaoming Zhou. 2018. "Evaluation of Collection-6 MODIS Land Surface Temperature Product Using Multi-Year Ground Measurements in an Arid Area of Northwest China." *Remote Sensing* 10 (11): 1852. https://doi.org/10.3390/rs10111852.

Ma, Xu, Anlong, Alfredo Huete, Qiang Yu, Natalia Restrepo Coupe, Kevin Davies, Mark Broich, et al. 2013. "Spatial Patterns and Temporal Dynamics in Savanna Vegetation Phenology Across the North Australian Tropical Transect." *Remote Sensing of Environment* 139. Elsevier Inc.: 97–115. https://doi.org/10.1016/j.rse.2013.07.030.

Mao, K. B., Y. Ma, X. L. Tan, X. Y. Shen, G. Liu, Z. L. Li, J. M. Chen, and L. Xia. 2017. "Global Surface Temperature Change Analysis Based on MODIS Data in Recent Twelve Years." *Advances in Space Research* 59 (2): 503–512. https://doi.org/10.1016/j.asr.2016.11.007.

Maselli, Fabio. 2004. "Monitoring Forest Conditions in a Protected Mediterranean Coastal Area by the Analysis of Multiyear NDVI data." *Remote Sensing of Environment* 89 (4): 423–433. https://doi.org/10.1016/j.rse.2003.10.020.

Mildrexler, David J., Maosheng Zhao, and Steven W. Running. 2011. "A Global Comparison Between Station Air Temperatures and MODIS Land Surface Temperatures Reveals the Cooling Role of Forests." *Journal of Geophysical Research: Biogeosciences* 116 (3): 1–15. https://doi.org/10.1029/2010JG001486.

Mishra, Niti B., and Gargi Chaudhuri. 2015. "Spatio-Temporal Analysis of Trends in Seasonal Vegetation Productivity Across Uttarakhand, Indian Himalayas, 2000–2014." *Applied Geography* 56. Elsevier Ltd: 29–41. https://doi.org/10.1016/j.apgeo.2014.10.007.

Mishra, Niti B., Kelley A. Crews, Neeti Neeti, Thoralf Meyer, and Kenneth R. Young. 2015. "MODIS Derived Vegetation Greenness

Trends in African Savanna: Deconstructing and Localizing the Role of Changing Moisture Availability, Fire Regime and Anthropogenic Impact." *Remote Sensing of Environment* 169. Elsevier Inc.: 192–204. https://doi.org/10.1016/j.rse.2015.08.008.

Mishra, Niti B., and Kumar P. Mainali. 2017. "Greening and Browning of the Himalaya: Spatial Patterns and the Role of Climatic Change and Human Drivers." *Science of The Total Environment* 587–588. Elsevier B.V: 326–339. https://doi.org/10.1016/j.scitotenv.2017.02.156.

Mohanty, Binayak P., Venkat Lakshmi, and Carsten Montzka. 2013. "Remote Sensing for Vadose Zone Hydrology – A Synthesis From the Vantage Point." *Vadose Zone Journal* 12: 1–6. https://doi.org/10.2136/vzj2013.07.0128.

Monteith, J. L. 1964. *Evaporation and Environment. The State and Movement of Water in Living Organisms.* Symposium of the Society of Experimental Biology, Vol. 19, 205–234. Cambridge: Cambridge University Press.

Monteith, J. L. 1965. "Evaporation and Environment." *Symposia of the Society for Experimental Biology* 19: 205–224.

Mu, Qiaozhen, Faith Ann Heinsch, Maosheng Zhao, and Steven W. Running. 2007. "Development of a Global Evapotranspiration Algorithm Based on MODIS and Global Meteorology Data." *Remote Sensing of Environment* 111: 519–536. https://doi.org/10.1016/j.rse.2007.04.015.

Mu, Qiaozhen, Maosheng Zhao, and Steven W. Running. 2011. "Improvements to a MODIS Global Terrestrial Evapotranspiration Algorithm." *Remote Sensing of Environment* 115 (8). Elsevier Inc.: 1781–1800. https://doi.org/10.1016/j.rse.2011.02.019.

Parmesan, Camille, and Gary Yohe. 2003. "A Globally Coherent Fingerprint of Climate Change Impacts Across Natural Systems." *Nature* 421 (January). Macmillian Magazines Ltd.: 37. https://doi.org/10.1038/nature01286.

Piao, Shilong, Xuhui Wang, Philippe Ciais, Biao Zhu, Tao Wang, and Jie Liu. 2011. "Changes in Satellite-Derived Vegetation Growth Trend in Temperate and Boreal Eurasia From 1982 to 2006." *Global Change Biology* 17 (10): 3228–3239. https://doi.org/10.1111/j.1365-2486.2011.02419.x.

Pielke, Roger A., Christopher A. Davey, Dev Niyogi, Souleymane Fall, Jesse Steinweg-Woods, Ken Hubbard, Xiaomao Lin, et al. 2007. "Unresolved Issues With the Assessment of Multidecadal Global Land Surface Temperature Trends." *Journal of Geophysical Research Atmospheres* 112 (24): 1–26. https://doi.org/10.1029/2006JD008229.

Qin, Zhihao, Giorgio Dall'Olmo, Arnon Karnieli, and Pedro Berliner. 2001. "Derivation of Split Window Algorithm and Its Sensitivity Analysis for Retrieving Land Surface Temperature From NOAA-Advanced Very High Resolution Radiometer Data." *Journal of Geophysical Research Atmospheres* 106 (D19): 22655–22670. https://doi.org/10.1029/2000JD900452.

Qu, Michael, Joe Wan, and Xianjun Hao. 2014. "Analysis of Diurnal Air Temperature Range Change in the Continental United States." *Weather and Climate Extremes* 4. Elsevier: 86–95. https://doi.org/10.1016/j.wace.2014.05.002.

Quéré, Le, Michael R. Raupach, Josep G. Canadell, Gregg Marland, Laurent Bopp, Philippe Ciais, Thomas J. Conway, et al. 2009. "Trends in the Sources and Sinks of Carbon Dioxide." *Nature Geoscience* 2 (12): 831–836. https://doi.org/10.1038/ngeo689.

Running, Steven W., Qiaozhen Mu, Maosheng Zhao, and Alvaro Moreno. 2019. "User's Guide MODIS Global Terrestrial Evapotranspiration (ET) Product NASA Earth Observing System MODIS Land Algorithm (For Collection 6)." *NASA's Goddard Space Flight Center.* https://modis-land.gsfc.nasa.gov/pdf/MOD16UsersGuideV2.022019.pdf.

Sun, Donglian, Rachel T. Pinker, and Menas Kafatos. 2006. "Diurnal Temperature Range Over the United States: A Satellite View." *Geophysical Research Letters* 33 (5): 2–5. https://doi.org/10.1029/2005GL024780.

Tierney, L, A. J. Rossini, Na Li, and H. Sevcikova. 2018. "Package 'Snow'. Simple Network of Workstations Version 0.4-3." https://CRAN.R-project.org/package=snow.

Urqueta, Harry, Jorge Jódar, Christian Herrera, Hans-G. Wilke, Agustín Medina, Javier Urrutia, Emilio Custodio, and Jazna Rodríguez. 2018. "Land Surface Temperature as an Indicator of the Unsaturated Zone Thickness: A Remote Sensing Approach in the Atacama Desert." *Science of The Total Environment* 612. Elsevier B.V.: 1234–1248. https://doi.org/10.1016/j.scitotenv.2017.08.305.

Verbesselt, Jan, and Martin Herold. 2012. "Near Real-Time Disturbance Detection Using Satellite Image Time Series: Drought Detection in Somalia." *Remote Sensing of Environment* 123: 98–108. https://doi.org/10.1016/j.rse.2012.02.022.

Verbesselt, Jan, Rob Hyndman, and Darius Culvenor. 2010. "Phenological Change Detection While Accounting for Abrupt and Gradual Trends in Satellite Image Time Series." *Remote Sensing of Environment* 114 (12): 2970–2980. https://doi.org/10.1016/j.rse.2010.08.003.

Verbesselt, Jan, Rob Hyndman, Glenn Newnham, and Darius Culvenor. 2010. "Detecting Trend and Seasonal Changes in Satellite Image Time Series." *Remote Sensing of Environment* 114: 106–115. https://doi.org/10.1016/j.rse.2009.08.014.

Wan, Zhengming. 2013. "MODIS Land Surface Temperature Products Users' Guide." *ERI*, University of California, Santa Barbara https://lpdaac.usgs.gov/documents/118/MOD11_User_Guide_V6.pdf.

Wan, Zhengming, and Z.-L. Li. 2008. "Radiance-Based Validation of the V5 MODIS Land-Surface Temperature Product." *International Journal of Remote Sensing* 29 (17–18). Taylor & Francis: 5373–5395. https://doi.org/10.1080/01431160802036565.

Wan, Zhengming, Y Zhang, Y.Q Zhang, and Z.-L Li. 2002. "Validation of the Land-Surface Temperature Products Retrieved From Moderate Resolution Imaging Spectroradiometer Data." *Remote Sensing of Environment* 83: 163–180.Wang, Wenhui, Shunlin Liang, and Tilden Meyers. 2008. "Validating MODIS Land Surface Temperature Products Using Long-Term Nighttime Ground Measurements." *Remote Sensing of Environment* 112 (3): 623–635. https://doi.org/10.1016/j.rse.2007.05.024.

Xu, Guang, Huifang Zhang, Baozhang Chen, Hairong Zhang, John L. Innes, Guangyu Wang, Jianwu Yan, Yonghong Zheng, Zaichun Zhu, and Ranga B. Myneni. 2014. "Changes in Vegetation Growth Dynamics and Relations With Climate Over China's Landmass from 1982 to 2011." *Remote Sensing* 6 (4): 3263–3283. https://doi.org/10.3390/rs6043263.

Yin, Gang, Zengyun Hu, Xi Chen, and Tashpolat Tiyip. 2016. "Vegetation Dynamics and Its Response to Climate Change in Central Asia." *Journal of Arid Land* 8 (3): 375–388. https://doi.org/10.1007/s40333-016-0043-6.

Zhou, Decheng, Liangxia Zhang, Dan Li, Dian Huang, and Chao Zhu. 2016. "Climate – Vegetation Control on the Diurnal and Seasonal Variations of Surface Urban Heat Islands in China." *Environmental Research Letters* 11 (7). IOP Publishing: 074009. https://doi.org/10.1088/1748-9326/11/7/074009.

## CHAPTER 3

Environmental Applications of Low Resolution Remotely Sensed Data

Bin Fang, Maria A. Banti, Venkat Lakshmi, and Alexandra Gemitzi

3.1 THE GRACE MISSION: GROUNDWATER ASSESSMENTS

GRACE stands for Gravity Recovery Climate Experience and corresponds to a joint mission of NASA and the German Aerospace Center. Twin satellites, i.e. the GRACE spacecrafts, were put in polar orbit on March 2002 and operated until October 2017. They were flying 220 kilometers apart at an altitude of 500 kilometers above Earth's surface. Very precise measurements of the distance between the two spacecrafts were obtained by means of GPS and a microwave ranging system that measures the variations of the separation distance of the satellites to within one micron (NASA-JPL

2019). Those measurements are converted to mass concentrations, mapping thus Earth's gravity field in time and space. Detected variations in the gravity field of Earth include changes related to surface and deep ocean currents, terrestrial total water storage changes, and exchanges of mass between ice sheets and glaciers and the ocean, as well as mass variations within Earth. Variations in the mass of atmosphere are removed along with ocean motions, recovering thus the hydrological and oceanographic signals (Wahr, Molenaar, and Bryan 1998). Available data comprise separate land and ocean grids at one degree (~110 km) to half degree (~50 km) spatial resolutions and are released as monthly changes in equivalent water thickness in cm relative to a time-mean baseline, i.e. 2004–2009.

GRACE Follow-On (GRACE-FO) is the mission currently continuing GRACE, with almost identical hardware, launched in May 2018. From 2019 it has started releasing data. Three different processing and distribution centers of GRACE and GRACE-FO data currently operate, i.e. Jet Propulsion Laboratory (JPL), Center for Space Research at University of Texas, Austin (CSR), and GeoforschungsZentrum Potsdam (GFZ). Parameter choices for converting distance observations to gravity field variations can be based on different solution strategies that are possible, and solutions from GFZ, CSR, and JPL explore those solution strategies differently. Recent works have shown that a simple average of those three solutions (ensemble mean) minimizes uncertainty and reduces noise in the solution field (Sakumura, Bettadpur, and Bruinsma 2014). Therefore users are advised to follow this approach (NASA-JPL 2019)

Applications of GRACE data are numerous and cover a wide area of environmental aspects. A key target application of GRACE is the monitoring of groundwater systems, and most of the numerous papers published during the last two decades concerning GRACE demonstrate the usefulness of gravitational observations in groundwater assessment. A significant depletion in groundwaters of northwest India from 2002 to 2008 was assessed in Rodell,

Velicogna, and Famiglietti (2009), whereas an overestimation of aquifer depletion in northeast India has been reported in Long et al. (2016). This latter work highlights uncertainties in GRACE data and the need for reliable ground information to refine the spatial patterns of GRACE signals. A methodology for estimation of GRACE data based Renewable Groundwater Stress (RGS) was published in Richey et al. (2015) and highlighted basins around the globe that may be vulnerable to higher levels of groundwater stress with driving factors mainly being the intensification of agriculture and population growth. An application of the same methodology in Greece (Gemitzi and Lakshmi 2018b) demonstrated that GRACE data may well be used along with ground information to determine RGS regimes even in aquifers with limited geographical extent. The use of ground monitoring measurements to downscale GRACE data for predictions of groundwater level changes was demonstrated in (Sun 2013) whereas Gemitzi and Lakshmi (2018a) developed a methodology to estimate groundwater abstractions at the aquifer scale using GRACE data and auxiliary ground information.

Drought monitoring is another crucial issue where GRACE has been found to contribute. Houborg et al. (2012) developed GRACE based drought indicators to identify drought conditions in North America and highlighted the improved potential for drought monitoring using GRACE observations.

The usefulness of GRACE data in assessing flood potential was demonstrated in Reager, Thomas, and Famiglietti (2014), where a relationship between measured river discharges in the Missouri River and GRACE derived total water storage in the whole basin was established.

GRACE observations are also useful in monitoring changes in large reservoirs. Wang et al. (2011) demonstrated how water storage changes in the Three Gorges reservoir in China are captured by GRACE. The crucial role of GRACE in detecting ice mass changes has been depicted by many publications. Velicogna, Sutterley, and Broeke (2014) determined the regional acceleration of ice mass losses in Greenland and Antarctica. Sea level rise monitoring can also be supported by GRACE observations. Reager et al. (2016) showed that there was a slowdown of sea level rise during 2002 to 2014, attributed to increased storage of water in land.

The GRACE mission supports the observation of co-seismic and post-seismic changes in gravitational field related to large earthquakes (>7.5 Richter) as well as the detections of post-seismic relaxation useful to study properties of Earth's crust and upper mantle (NASA-JPL 2019). A major limitation arises, however, from the coarse spatial resolution of GRACE data.

Regarding GRACE and GRACE-FO data access many different options are available outlined in https://podaac.jpl.nasa.gov/dataaccess. A handy way to access GRACE along with numerous other remotely sensed products is the PO.DAAC's (Physical Oceanography Data Active Archive Center) WebDAV interface which allows users to connect to PO.DAAC and download data as if it were a local drive on their computer (https://podaac-tools.jpl.nasa.gov/drive/help). Users are strongly advised to read the related data product handbook before they proceed to application.

The numerous publications of GRACE related research demonstrate the potential of such missions to provide state of the art monitoring. It is certain that advances in remote sensing and information technologies in the near future will provide improved information. GRACE based information does not inherit limitations due to e.g. cloud cover, but users should keep in mind that uncertainly in GRACE estimates is large mainly due to uncertainties in the processing parameters. Another source of uncertainty is the native spatial resolution of ~330 km, whereas the Level-3 gridded data are provided at one degree to half degree. Therefore, in areas smaller than 100,000 km², the noise to signal ratio provides poor results. Researchers, however, should keep in mind that downscaling of GRACE data using in situ monitoring information may provide very useful information for local scale basins as well.

3.2 SMAP AND SMOS MISSIONS FOR SOIL MOISTURE DETERMINATION

Introduction

Soil moisture (SM) observations provide important information for studying the exchange of water and energy between the land and the atmosphere and serve as a key input variable in numerical models for weather forecasting, ecology, and agriculture applications as well as prediction of hydrological extremes such as floods and droughts. As compared to point scale SM from ground-based sensors, satellite observations (in particular microwave-based remote sensing) provide routine and almost continuous global coverage (Jackson 1993; Schmugge and Jackson 1994; Lakshmi 2013). During the past decade, a number of active and passive microwave satellites have been launched that are capable of measuring surface moisture at global scale. These include Advanced Microwave Scanning Radiometer for the Earth Observing System (AMSR-E), Advanced Microwave Scanning Radiometer 2 (AMSR2), Advanced SCATterometer (ASCAT), Aquarius, Soil Moisture and Ocean Salinity (SMOS), and Soil Moisture Active/ Passive (SMAP).

The spatial resolutions of the currently available passive microwave SM products are determined by the antenna dimensions and are on the order of tens of kilometers (McCabe et al. 2017). Such spatial resolution does not satisfy the demands of many hydrological or agricultural studies that require characterization of SM variability at higher scales. Recent research has examined the use of downscaling algorithms as a solution to the spatial resolution problem (Merlin et al. 2015; Peng, Niesel, and Loew 2015; Sabaghy et al. 2018).

The SMAP satellite is an L-band mission (radiometer centered at 1.41 GHz) that operates in a near polar sun synchronous orbit. It has a revisit time of two to three days and observes Earth at 6 a.m./6 p.m. The SM estimates represent the top 0–5 cm of the soil layer with a threshold of vegetation water content ≤ 5 kg/m^2 (Chan et al. 2016; Chan et al. 2017; Colliander et al. 2017; McNairn et al. 2014).

The enhanced Level-2 half orbit 9 km SM product (SPL2SMP_E) can be acquired from NSIDC (National Snow and Ice Data Center) at https://nsidc.org/data/SPL2SMP_E. The SMAP SM retrievals are solely from the L-band radiometer T_B data using the Single Channel Algorithm (Jackson 1993) at the native spatial resolution of 33–40 km and enhanced to 9 km EASE (Equal Area Scalable Earth) grid using the approach developed by Chan et al. (2017), which is based on interpolating antenna temperature data in the original SMAP Level-1B T_B data, by using Backus-Gilbert optimal interpolation technique.

Apart from NASA's SMAP mission, the SMOS satellite, which constitutes a mission concept designed by the European Space Agency (ESA), was launched on November 2009 (Knipper et al. 2017). The SMOS mission is devoted to the global measurement of soil moisture, as well as to the determination of the ocean salinity state, on a 6 a.m./6 p.m. orbit with a time of revisit shorter than three days (Kerr et al. 2010). The satellite's payload is an L-Band 2-D interferometric radiometer which operates in a frequency of 1.4–1.42 GHz (Piles et al. 2011; Kerr et al. 2010). These requirements enable the acquisition of soil moisture estimates from the top 5 cm of the soil column (Merlin et al. 2012) at a spatial resolution of ~50 km (Piles et al. 2011). Moreover, the SMOS mission targets on providing estimates of soil moisture within an accuracy of 0.04 m^3/m^3 , while the accuracy of the vegetation water content is evaluated to be 0.1 kg/m^2 (Kerr et al. 2010). As related to soil moisture, two levels of soil moisture products are provided by this mission either on a ~25 km Equal-Area Scalable Earth (EASE) grid or on a 15 km Icosahedral Snyder Equal Area (ISEA) 4H9 grid (European Space Agency 2017).

In this investigation, an improved SM downscaling algorithm is proposed focusing on the enhanced 9 km SMAP data, and the downscaling is done using satellite VIS/IR (Visible/Infrared) observations over the Contiguous United States (CONUS) domain. The algorithm originally exploited the previously well-studied Normalized Difference Vegetation Index (NDVI) modulated relationship between SM and surface temperature (Carlson 2007; Gillies, Kustas, and Humes 1997; Mallick, Bhattacharya, and Patel 2009; Minacapilli,

Iovino, and Blanda 2009; Lakshmi et al. 2001; Lakshmi et al. 2011). The algorithm is performed on the enhanced 9 km SMAP SM over 33 km domain, which is the native domain to the SMAP radiometer observations. It utilizes the high spatial resolution satellite surface temperature and VIS/IR-based vegetation index from MODerate resolution Imaging Spectroradiometer (MODIS) to produce a wide coverage and routinely available SM retrievals (Fang et al. 2013; Fang et al. 2018).

Methodology

The downscaling algorithm is based on the thermal inertia theory that wetter soils tend to have greater volumetric heat capacity and lower temperature change, and vice versa for dry soils. The assumption made is that the rate of change in surface SM is negatively correlated to the rate of change in surface temperature (Matsushima, Kimura, and Shinoda 2012). This is especially true during the summer months when the evapotranspiration rate is the dominant factor. Therefore, in the application of the algorithm, it is assumed that the temperature difference between 1:30 a.m./1:30 p.m., which are the two overpass times of Aqua MODIS, would correspond to the SMAP SM overpasses, i.e. 6 a.m. or 6 p.m.

It is assumed that the maximum diurnal surface temperature difference can be estimated by the difference of surface temperature between 1:30 p.m. (time of maximum surface temperature) and 1:30 a.m. (time of minimum surface temperature) (Fang et al. 2013). The relationship between the North American Land Data Assimilation System (NLDAS) derived surface skin temperature change ΔT_s and SM θ can be modeled using linear regression fit as described in Equation (3.1).

$$\theta(i,j) = a_0 + a_1 \Delta T_s(i,j) \tag{3.1}$$

For any NLDAS grid point (i, j) there are two equations that describe the Δ–θ relationship for the SMAP morning and evening overpasses SM $\theta^a(i,j)$ and $\theta^p(i,j)$, respectively. a_0 and a_1 are the best fit regression coefficients.

This linear regression model can be applied using NLDAS Noah model ΔT_s and θ estimates between 1981 and 2018. This step allows the development of reliable regression coefficients a_0 and a_1. Additionally, the $\theta - \Delta T_s$ relationship is modeled separately for each month as the relationship exhibits seasonal variation (Fang et al. 2018).

Furthermore, the LTDR NDVI data is used to classify the corresponding NLDAS $\theta - \Delta T_s$ relationship by grouping them at an interval of 0.1 over the 0–0.8 NDVI range. The relationship proposed in Equation (3.1) is then developed separately for each class and applied on 1 km MODIS land surface temperature (LST) grids which are included within the corresponding NLDAS grid's boundaries to calculate the 1 km SM. An assumption is made that the variance of the $\theta - \Delta T_s$ relationship within one NLDAS grid could be ignored.

The 1 km SM calculated from the $\theta - \Delta T_s$ model needs to be adjusted by the original 9 km SMAP SM. This is because: (1) The SMAP 9 km SM is retrieved from microwave radiometer L-band brightness temperature (T_B) observations, while the 1 km $\theta - \Delta T_s$ model SM output is calculated from MODIS LST, which is retrieved from spectroradiometer observations at VIS/IR bands. (2) The microwave radiometer represents SM at 0–5 cm layer, while the $\theta - \Delta T_s$ model SM outputs are calculated from MODIS LST and represent only a few mm from the surface. Therefore, in order to remove these discrepancies, the SMAP 9 km SM of morning/evening overpasses Θ were corrected using the difference between the SMAP 9 km SM and the average of all 1 km $\theta - \Delta T_s$ model output grids located in the 33 km contributing domain corresponding to that 9 km grid, which is the native spatial area to the SMAP radiometer T_B (Chan et al. 2017). This modification can effectively reduce the sharp edge effect which is most likely caused by the errors of NLDAS and MODIS data. This is the major improvement of the original downscaling algorithm developed by Fang et al. (2013). The equation for adjusted SM θ^c is written as:

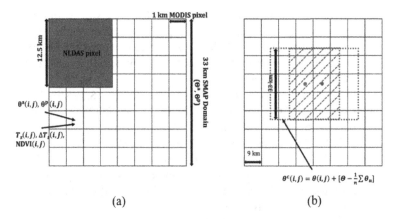

(a)   (b)

FIGURE 3.1   Conceptual diagram of (a) fitting and implementing the $\theta-\Delta T_s$ relationship equation from NLDAS model grid size (12.5 km) to MODIS pixel size (1 km); (b) performing the correction of 9 km SMAP SM using SM which is calculated from the $\theta-\Delta T_s$ model at 33 km domain. The red and blue boxes denote the two adjacent 33 km domains of the 9 km apart SMAP SM points.

$$\theta^c\left(i,j\right)=\theta\left(i,j\right)+\left[\Theta-\frac{1}{n}\sum\theta_n\right] \qquad (3.2)$$

Where i, j represent a 1 km pixel of $\theta-\Delta T_s$ model output SM and n is the number of 1 km pixels included by 9 km SMAP SM pixel. Θ is the 9 km SMAP SM estimates. The conceptual diagram of the model building, implementation, and downscaling is given in Figure 3.1.

REFERENCES

Carlson, Toby. 2007. "An Overview of the "Triangle Method" for Estimating Surface Evapotranspiration and Soil Moisture From Satellite Imagery." *Sensors* 7 (8): 1612–1629.

Chan, S. K., R. Bindlish, P. O. Neill, T. Jackson, E. Njoku, S. Dunbar, J. Chaubell, et al. 2017. "Development and Assessment of the SMAP Enhanced Passive Soil Moisture Product." *Remote Sensing of Environment* 204: 931–941. https://doi.org/10.1016/j.rse.2017.08.025.

Chan, Steven K., Rajat Bindlish, Peggy E. O. Neill, Eni Njoku, Tom Jackson, Andreas Colliander, et al. 2016. "Assessment of the SMAP Passive Soil Moisture Product." *IEEE Transactions on Geoscience and Remote Sensing* 54 (8): 4994–5007. https://doi.org/10.1109/TGRS.2016.2561938.

Colliander, A., T. J. Jackson, R. Bindlish, S. Chan, N. Das, S. B. Kim, M. H. Cosh, et al. 2017. "Validation of SMAP Surface Soil Moisture Products With Core Validation Sites." *Remote Sensing of Environment* 191: 215–231. https://doi.org/10.1016/j.rse.2017.01.021.

European Space Agency. 2017. "Smos Data Products." *European Space Agency*. https://earth.esa.int/web/guest/missions/esa-operational-eo-missions/smos/content/-/asset_publisher/t5Py/content/data-types-levels-formats-7631.

Fang, Bin, Venkataraman Lakshmi, Rajat Bindlish, and Thomas J. Jackson. 2018. "Downscaling of SMAP Soil Moisture Using Land Surface Temperature and Vegetation Data." *Vadose Zone Journal* 17 (1). https://doi.org/10.2136/vzj2017.11.0198.

Fang, Bin, Venkataraman Lakshmi, Rajat Bindlish, Thomas J. Jackson, Michael Cosh, and Jeffrey Basara. 2013. "Passive Microwave Soil Moisture Downscaling Using Vegetation Index and Skin Surface Temperature." *Vadose Zone Journal* 12 (3). https://doi.org/10.2136/vzj2013.05.0089.

Gemitzi, Alexandra, and Venkat Lakshmi. 2018a. "Estimating Groundwater Abstractions at the Aquifer Scale Using GRACE Observations." *Geosciences* 8: 1–14. https://doi.org/10.3390/geosciences8110419.

Gemitzi, Alexandra, and Venkat Lakshmi, 2018b. "Evaluating Renewable Groundwater Stress With GRACE Data in Greece." *Groundwater* 56: 501–514. https://doi.org/10.1111/gwat.12591.

Gillies, R. R., W. P. Kustas, and K. S. Humes. 1997. "A Verification of the "Triangle" Method for Obtaining Surface Soil Water Content and Energy Fluxes From Remote Measurements of the Normalized Difference Vegetation Index (NDVI) and Surface." *International Journal of Remote Sensing* 18 (15): 3145–3166. https://doi.org/10.1080/014311697217026.

Houborg, Rasmus, Matthew Rodell, Bailing Li, Rolf Reichle, and Benjamin F. Zaitchik. 2012. "Drought Indicators Based on Model-Assimilated Gravity Recovery and Climate Experiment (GRACE) Terrestrial Water Storage Observations." *Water Resources Research* 48: 1–17. https://doi.org/10.1029/2011WR011291.

Jackson, Thomas J. 1993. "III. Measuring Surface Soil Moisture Using Passive Microwave Remote Sensing." *Hydrological Processes* 7 (2): 139–152. https://doi.org/10.1002/hyp.3360070205.

Kerr, Yann H., Philippe Waldteufel, Jean Pierre Wigneron, Steven Delwart, François Cabot, Jacqueline Boutin, Maria José Escorihuela, et al. 2010. "The SMOS Mission: New Tool for Monitoring Key Elements of the Global Water Cycle." *Proceedings of the IEEE* 98 (5): 666–687. https://doi.org/10.1109/JPROC.2010.2043032.

Knipper, Kyle R., Terri S. Hogue, Kristie J. Franz, and Russell L. Scott. 2017. "Downscaling SMAP and SMOS Soil Moisture With Moderate-Resolution Imaging Spectroradiometer Visible and Infrared Products Over Southern Arizona." *Journal of Applied Remote Sensing* 11 (2): 026021. https://doi.org/10.1117/1.JRS.11.026021.

Lakshmi, Venkat. 2013. "Remote Sensing of Soil Moisture." *ISRN Soil Science.* Article ID 424178. https://doi.org/10.1155/2013/424178.

Lakshmi, Venkat, K. Czajkowski, R. Dubayah, and J. Susskind. 2001. "Land Surface Air Temperature Mapping Using TOVS and AVHRR." *International Journal of Remote Sensing* 22 (4): 643–662. https://doi.org/10.1080/01431160050505900.Lakshmi, Venkat, Seungbum Hong, Eric E. Small, and Fei Chen. 2011. "The Influence of the Land Surface on Hydrometeorology and Ecology: New Advances From Modeling and Satellite Remote Sensing." *Hydrology Research* 42 (2–3): 95–112. https://doi.org/10.2166/nh.2011.071.

Long, Di, Xi Chen, Bridget R. Scanlon, Yoshihide Wada, Yang Hong, Vijay P. Singh, Yaning Chen, Cunguang Wang, Zhongying Han, and Wenting Yang. 2016. "Have GRACE Satellites Overestimated Groundwater Depletion in the Northwest India Aquifer?" *Nature Publishing Group*, no. April. Nature Publishing Group: 1–11. https://doi.org/10.1038/srep24398.

Mallick, Kaniska, Bimal K. Bhattacharya, and N. K. Patel. 2009. "Estimating Volumetric Surface Moisture Content for Cropped Soils Using a Soil Wetness Index Based on Surface Temperature and NDVI." *Agricultural and Forest Meteorology* 149 (8): 1327–1342. https://doi.org/10.1016/j.agrformet.2009.03.004.

Matsushima, D., R. Kimura, and M. Shinoda. 2012. "Soil Moisture Estimation Using Thermal Inertia: Potential and Sensitivity to Data Conditions." *Journal of Hydrometeorology* 13 (2): 638–648. https://doi.org/10.1175/JHM-D-10-05024.1.

McCabe, Matthew F., Matthew Rodell, Douglas E. Alsdorf, Diego G. Miralles, Remko Uijlenhoet, Wolfgang Wagner, Arko Lucieer, et al. 2017. "The Future of Earth Observation in Hydrology." *Hydrology and Earth System Sciences* 21 (7): 3879–3914. https://doi.org/10.5194/hess-21-3879-2017.

McNairn, Heather, Thomas J Jackson, Grant Wiseman, Stéphane Bélair, Aaron Berg, Paul Bullock, Andreas Colliander, et al. 2014. "The Soil Moisture Active Passive Validation Experiment 2012 (SMAPVEX12): Prelaunch Calibration and Validation of the SMAP Soil Moisture Algorithms." *IEEE Transactions on Geoscience and Remote Sensing* 53 (5): 2784–2801. https://doi.org/10.1109/TGRS.2014.2364913.

Merlin, Olivier, Yoann Malbéteau, Youness Notfi, Stefan Bacon, Salah Er-raki, Said Khabba, and Lionel Jarlan. 2015. "Performance Metrics for Soil Moisture Downscaling Methods: Application to DISPATCH Data in Central Morocco." *Remote Sensing* 7 (4): 3783–3807. https://doi.org/10.3390/rs70403783.

Merlin, Olivier, Christoph Rudiger, Ahmad Al Bitar, Philippe Richaume, Jeffrey P. Walker, and Yann H. Kerr. 2012. "Disaggregation of SMOS Soil Moisture in Southeastern Australia." *IEEE Transactions on Geoscience and Remote Sensing* 50 (5): 1556–1571. https://doi.org/10.1109/TGRS.2011.2175000.

Minacapilli, M., M. Iovino, and F. Blanda. 2009. "High Resolution Remote Estimation of Soil Surface Water Content by a Thermal Inertia Approach." *Journal of Hydrology* 379 (3–4): 229–238. https://doi.org/10.1016/j.jhydrol.2009.09.055.

NASA Jet Propulsion Laboratory. 2019. *GRACE D-103133 Gravity Recovery and Climate Experiment Level-3 Data Product User Handbook.* https://podaac-tools.jpl.nasa.gov/drive/files/allData/gracefo/docs/GRACE-FO_L3_Handbook_JPL-D-103133_20190327.pdf.

Peng, J., J. Niesel, and A. Loew. 2015. "Evaluation of Soil Moisture Downscaling Using a Simple Thermal-Based Proxy – The REMEDHUS Network (Spain) Example." *Hydrology and Earth System Sciences* 19 (12): 4765–4782. https://doi.org/10.5194/hess-19-4765-2015.

Piles, María, Adriano Camps, Mercè Vall-Llossera, Ignasi Corbella, Rocco Panciera, Christoph Rudiger, Yann H. Kerr, and Jeffrey Walker. 2011. "Downscaling SMOS-Derived Soil Moisture Using MODIS Visible/Infrared Data." *IEEE Transactions on Geoscience and Remote Sensing* 49 (9): 3156–3166. https://doi.org/10.1109/TGRS.2011.2120615.

Reager, J. T., B. F. Thomas, and J. S. Famiglietti. 2014. "River Basin Flood Potential Inferred Using GRACE Gravity Observations at Several Months Lead Time." *Nature Geoscience*, no. August: 588–592. https://doi.org/10.1038/NGEO2203.

Reager, J. T., A. S. Gardner, J. S. Famiglietti, D. N. Wiese, A. Eicker, and M.-H. Lo. 2016. "A Decade of Sea Level Rise Slowed by Climate-

Driven Hydrology." *Science* 351: 699–703.Richey, A. S., B. F. Thomas, M-H Lo, J. T. Reager, J. S. Famiglietti, K. Voss, S. Swenson, and M. Rodell. 2015. "Quantifying Renewable Groundwater Stress With GRACE." *Water Resources Research* 51: 5217–5238. https://doi.org/10.1002/2015WR017349.

Rodell, Matthew, Isabella Velicogna, and James S. Famiglietti. 2009. "Satellite-Based Estimates of Groundwater Depletion in India." *Nature* 460 (7258). Nature Publishing Group: 999–1002. https://doi.org/10.1038/nature08238.

Sabaghy, Sabah, J. P. Walker, Luigi J. Renzullo, and Thomas J. Jackson. 2018. "Spatially Enhanced Passive Microwave Derived Soil Moisture: Capabilities and Opportunities." *Remote Sensing of Environment* 209: 551–580. https://doi.org/10.1016/j.rse.2018.02.065.

Sakumura, C., S. Bettadpur, and S. Bruinsma. 2014. "Ensemble Prediction and Intercomparison Analysis of GRACE Time-Variable Gravity Field Models." *Geophysical Research Letters*, 1389–1397. https://doi.org/10.1002/2013GL058632.1.

Schmugge, T., and T. J. Jackson. 1994. "Mapping Surface Soil Moisture With Microwave Radiometers." *Meteorology and Atmospheric Physics* 54 (1–4): 213–223. https://doi.org/10.1007/BF01030061.

Sun, A. Y. 2013. "Predicting Groundwater Level Changes Using GRACE Data." *Water Resources Research* 49: 5900–5912. https://doi.org/10.1002/wrcr.20421.

Velicogna, I., T. C. Sutterley, and M. R. Van Den Broeke. 2014. "Regional Acceleration in Ice Mass Loss From Greenland and Antarctica Using GRACE Time-Variable Gravity Data." no. January 2003: 8130–8137. https://doi.org/10.1002/2014GL061052.Received.

Wahr, J., M. Molenaar, and F. Bryan. 1998. "Time-Variability of the Earth's Gravity Field: Hydrological and Oceanic Effects and their Possible Detection Using GRACE." *Journal of Geophysical Research: Solid Earth* 103 (B12): 30205–30230. https://doi.org/10.1029/98JB02844.

Wang, X., C. De Linage, J. Famiglietti, and C. S. Zender. 2011. "Gravity Recovery and Climate Experiment (GRACE) Detection of Water Storage Changes in the Three Gorges Reservoir of China and Comparison With In Situ Measurements." *Water Resources Research* 47 (12): 1–13. https://doi.org/10.1029/2011WR010534.

## CHAPTER 4

Processing Remotely Sensed Data

Ibrahim N. Mohammed, John Bolten, Bin Fang, Maria A. Banti, and Venkat Lakshmi

4.1 DOWNSCALING OF REMOTELY SENSED DATA

Soil Moisture

The available downscaling algorithms can be classified as either methods using data from satellite observations, information from the other related land surface variables such as vegetation and soil properties, or model based (land surface model [LSM] or statistical-based approaches) (Peng et al. 2017). Some of the methods used to downscale coarse spatial resolution passive microwave soil moisture (SM) retrievals rely on the use of higher resolution (1) radar observations (Bolten, Lakshmi, and Njoku 2003; Das et al. 2014; Narayan, Lakshmi, and Njoku 2004; Narayan et al. 2006; Narayan et al. 2008; Kerr et al. 2001; Hornáček et al. 2012; Li et al. 2018; Jagdhuber et al. 2019), (2) satellite VIS/IR (Visible/Infrared) observations (Merlin et al. 2008, 2010; Fang et al. 2013; Fang and

Lakshmi 2014; Colliander et al. 2017a, 2017b; Mishra et al. 2018), or (3) LSM derived geophysical variables at high spatial resolution (Piles et al. 2011; Peng, Niesel, and Loew 2015a; Peng et al. 2015b). There are other studies that have developed downscaling models by using advanced mathematical approaches. For example, a random forest algorithm derived downscaling approach was proposed by Zhao et al. (2018), where the SMAP SM was downscaled using MODerate resolution Imaging Spectroradiometer (MODIS) products such as land surface temperature (LST) and vegetation index. Ajami and Sharma (2018) developed a hybrid downscaling method that applied the first order autoregressive model on high resolution hydrological model SM simulations to downscale LSM SM output. Another SMAP downscaling method based on an ensemble learning method and atmospheric and geophysical data was developed by Abbaszadeh, Moradkhani, and Zhan (2019). The observed area by the satellite is large, and to deal with the inherent heterogeneity Gaur and Mohanty (2019) introduced a scaling factor that accounts for the geophysical heterogeneity and wetness variability within the satellite footprint, and the approach was tested using SMOS SM observations. The Gaussian-Mixture non-stationary hidden Markov model was introduced to build a downscaling model for Advanced Microwave Scanning Radiometer 2 (AMSR2) SM (Kwon, Kwon, and Han 2017). The geographically weighted regression area to area kriging method was applied to downscaled AMSR2 SM (Jin et al. 2017).

Land Surface Temperature

The approaches for downscaling LST mostly build the relationship models between LST and other related biophysical parameters which are available at high spatial resolution (Hutengs and Vohland 2016). The biophysical parameters can be derived from remote sensing observations at visible/near-infrared bands, topography, and land surface modeled output data. Basically, the linear regression fit model is first implemented between LST and these parameters at coarse resolution and then applied on the fine resolution parameters. For example, Chen et al. (2014) applied Thin Plate Spline (TPS) interpolation approach, which is an effective technique for downscaling to combine with the Thermal sHARPening algorithm (TsHARP), and tested the method on various landscapes. A high resolution urban thermal sharpener (HUTS) method was developed by fitting the relationship between LST, Normalized Difference Vegetation Index (NDVI), and surface albedo and applied to downscale thermal infrared (TIR) data at 90 m resolution (Dominguez et al. 2011). A new method using data mining sharpener (DMS) technique is developed by building regression trees between TIR band brightness temperatures and shortwave spectral reflectances based on intrinsic sample characteristics (Gao, Kustas, and Anderson 2012). On another hand, previous research has found that the Vegetation Indices (VI) are highly related to thermal sharpening especially in agricultural purposes, and consequently, the TsHARP method is the most widely used to relate VI to LST. For example, an algorithm for downscaling LST data to 1 km resolution was developed by combining visible to thermal infrared information from daily MODIS 1 km observations and GOES (Geostationary Operational Environmental Satellite) 10 imager. (Inamdar et al. 2009). Jeganathan et al. (2011) tested and discussed the performance of the TsHARP model in the northern part of India by analyzing the five variants of the model using the Advanced Space-borne Thermal Emission Reflection Radiometer (ASTER) thermal data at different aggregated spatial resolutions. In recent years, more and more studies have tried to conduct non-linear regression methods along with more related geophysical parameters for downscaling LST, including using Artificial Neural Networks (ANN) method (Yang et al. 2010, Yang et al. 2011) and regression trees method (Gao, Kustas, and Anderson 2012)

Precipitation

The main approach to downscale remote sensing precipitation products including Tropical Rainfall Measuring Mission (TRMM) and Global Precipitation Measurement (GPM) is based on the relationships among precipitation and other land surface variables. The statistical model is built at coarse spatial resolution and then applied to fine resolution land surface variables for downscaling precipitation. For example, various studies used spatial and statistical analyses to downscale TRMM precipitation, including: Ezzine et al. (2017) used three predictors normalized difference water index (NDWI), elevation, and distance from sea to downscale TRMM data.

Ulloa et al. (2017) developed a two-step downscaling approach by integrating kriging regression model and in situ measurements. An exponential relationship between NDVI and precipitation were applied to downscaled TRMM data over the Iberian Peninsula (Immerzeel, Rutten, and Droogers 2009). NDVI and DEM (Digital Elevation Model) were used along with multiple linear regression (MLR) model to downscale TRMM data from 25 km to 1 km (Jia et al. 2011). Moreover, there were methods proposed using a geographically weighted regression (GWR) model to downscale precipitation data (Li et al. 2010; An et al. 2016). Additionally, Zhan et al. (2018) applied both GWR and MLR methods to downscale Global Precipitation Measurement (GPM) data in mountainous area. The EVI was used instead of the NDVI for downscaling TRMM data (Chen et al. 2019).

4.2 DATA FUSION TECHNIQUES: STUDYING FLOODS AND THEIR IMPACTS

Improving hydrological modeling and assessment techniques with the use of remote sensing data is a field of study that is progressing as a result of the recent availability of remote sensing observations and improving accuracy, fidelity, and novel advances that enable observations at finer spatial and temporal resolutions. The benefit of satellite-based Earth observations data for flood modeling in regions that experience poor spatial in situ earth observations data representation has been highlighted in recent studies conducted at the Lower Mekong River Basin (LMRB) (Mohammed et al. 2018a, 2018b).

The Mohammed et al. (2018a) study assessed the performance of satellite-based data products, TRMM and GPM, in estimating precipitation and streamflow over the Lower Mekong River Basin. Mohammed et al.'s study noticed that the quality of the satellite-based remote sensing precipitation data, especially in the southern part of the Lower Mekong River Basin (i.e., close to the delta), was better than elsewhere in the basin (Figure 4.1).

The ability of the modified hydrological model to represent the variability of the observed discharge at multiple sites along the Lower Mekong River when driven by satellite-based earth observation data corroborates the role of quality climate forcing as one of the main determinants in hydrologic modeling. Mohammed et al. (2018a) discharge results obtained from satellite-based remote sensing data were able to explain more than 91% of the variance observed in the monthly discharge during the calibration years (i.e., the Nash–Sutcliffe Efficiency [NSE] varied from 0.91 to 0.96 from sub-basin 1 to 6). The LMRB model performance when driven with in situ data was able to explain from 68% to 91% of the variance observed in the monthly discharge during the calibration years.

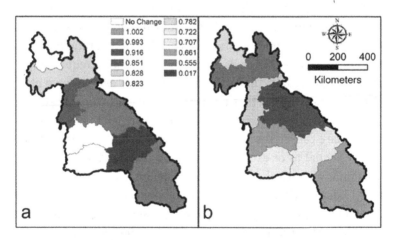

FIGURE 4.1 Precipitation data adjustment layout for (a) in situ station, and (b) satellite-based remote sensing (RS) model input excerpted from Mohammed et al. (2018a).

, The seasonal variations in the Mekong River's flow typically follow a dry season (December to May) and a wet season or a flood season (July to October). Lower Mekong flow records at the June, July, August, and September months are usually very high compared with other flow records during the year. Previous work has found that the main causes for flow alterations in the Mekong River were attributed to human activities (i.e., dam constructions) and climate change effects (Hoang et al. 2016; Wang et al. 2017). Mohammed et al. (2018b) formulated and evaluated hypothetical experiments for the Upper Mekong River Basin runoff yield increase and decrease scenarios to assess the impacts associated on the Lower Mekong River Basin flow seasonal variations. The Mohammed et al. (2018b) study suggests that the Lower Mekong flow variability and predictability conditions will be directly affected in terms of increased flow variability and decreased flow predictability due to changing input flow releases from the Upper Mekong River Basin.

Mohammed et al. (2018b) showed that upon changing the inflow input from upstream (i.e., Upper Mekong), the downstream streamflow predictability (Colwell 1974) has been reduced further. Mohammed et al. (2018b) studied high flow disturbance variables such as flood duration. Generally, flood durations within the Lower Mekong River Basin are long and more frequent especially at Mukdahan, Thailand, and southward. Mohammed et al. (2018b) results reveal that releasing more flows from the Upper Mekong River Basin (i.e., 30% increase) in 2011, for instance, would have caused the streamflow gauge to record 16 days of discharges equal to or exceeding the flood threshold at Mukdahan, Thailand. Historical records at Mukdahan, Thailand, indicated that 13 days with discharges equal to or exceeding a threshold of 30,400 m³/sec (equivalent to 12.6 m as a stage height) have been recorded. Mohammed et al. (2018b) also showed that releasing more flows from the Upper Mekong River Basin would also affect the frequency of flood occurrences. It is good to mention here that various hydrological time series and remote sensing data products needed to model and understand the Lower Mekong streamflow variability are available at Mohammed et al. (2018). In summary, Mohammed et al.'s works on the Lower Mekong floods suggested that flow releases increase from the Upper Mekong River Basin would mean more flooded days as well as higher frequency of flood occurrences specially at Mukdahan (Thailand), Pakse (Laos), and Kratie (Cambodia). This is an alarming finding since the fate of many people and properties will be at stake.

Another example of data fusion is the recent work by Oddo, Ahamed, and Bolten (2018). Oddo et al. blended an operational near-real-time satellite-based flood inundation system with a flood hazard model to provide a socioeconomic damage assessment model. In Oddo et al.'s study, flood impacts from the 2011 Southeast Asian flood were provided by a novel approach of estimating flood inundation depth with calibrated estimated damages based on land cover type.

REFERENCES

Abbaszadeh, Peyman, Hamid Moradkhani, and Xiwu Zhan. 2019. "Downscaling SMAP Radiometer Soil Moisture Over the CONUS Using an Ensemble Learning Method." *Water Resources Research* 55 (1): 324–344. https://doi.org/10.1029/2018WR023354.

Ajami, Hoori, and Ashish Sharma. 2018. "Disaggregating Soil Moisture to Finer Spatial Resolutions: A Comparison of Alternatives." *Water Resources Research* 54 (11): 9456–9483. https://doi.org/10.1029/2018WR022575.

An, Kyoung Jin, Sang Woo Lee, Soon Jin Hwang, Se Rin Park, and Sun Ah Hwang. 2016. "Exploring the Non-Stationary Effects of Forests and Developed Land Within Watersheds on Biological Indicators of Streams Using Geographically-Weighted Regression." *Water* 8 (4). https://doi.org/10.3390/w8040120.

Bolten, John D., Venkataraman Lakshmi, and Eni G. Njoku. 2003. "Soil Moisture Retrieval Using the Passive/Active L-and S-Band Radar/Radiometer." *IEEE Transactions on Geoscience and Remote Sensing* 41 (12): 2792–2801. https://doi.org/10.1109/TGRS.2003.815401.

Chen, Shaodan, Liping Zhang, Dunxian She, and Jie Chen. 2019. "Spatial Downscaling of Tropical Rainfall Measuring Mission (TRMM)

Annual and Monthly Precipitation Data Over the Middle and Lower Reaches of the Yangtze River Basin, China." *Water* 11 (3): 568. https://doi.org/10.3390/w11030568.

Chen, Xuehong, Wentao Li, Jin Chen, Yuhan Rao, and Yasushi Yamaguchi. 2014. "A Combination of TsHARP and Thin Plate Spline Interpolation for Spatial Sharpening of Thermal Imagery." *Remote Sensing* 6 (4): 2845–2863. https://doi.org/10.3390/rs6042845.

Colliander, A., T. J. Jackson, R. Bindlish, S. Chan, N. Das, S. B. Kim, M. H. Cosh, et al. 2017a. "Validation of SMAP Surface Soil Moisture Products With Core Validation Sites." *Remote Sensing of Environment* 191: 215–231. https://doi.org/10.1016/j.rse.2017.01.021.

Colliander, Andreas, Joshua B. Fisher, Gregory Halverson, Olivier Merlin, Sidharth Misra, Rajat Bindlish, Thomas J. Jackson, and Simon Yueh. 2017b. "Spatial Downscaling of SMAP Soil Moisture Using MODIS Land Surface Temperature and NDVI during SMAPVEX15." *IEEE Geoscience and Remote Sensing Letters* 14 (11): 2107–2111. https://doi.org/10.1109/LGRS.2017.2753203.

Colwell, R. K. 1974. "Predictability, Constancy, and Contingency of Periodic Phenomena." *Ecology* 55: 1148–1153. https://doi.org/10.2307/1940366.

Das, Narendra Narayan, Dara Entekhabi, Eni G. Njoku, Jiancheng J. C. Shi, Joel T. Johnson, and Andreas Colliander. 2014. "Tests of the SMAP Combined Radar and Radiometer Algorithm Using Airborne Field Campaign Observations and Simulated Data." *IEEE Transactions on Geoscience and Remote Sensing* 52 (4): 2018–2028. https://doi.org/10.1109/TGRS.2013.2257605.

Dominguez, Anthony, Jan Kleissl, Jeffrey C. Luvall, and Douglas L. Rickman. 2011. "High-Resolution Urban Thermal Sharpener (HUTS)." *Remote Sensing of Environment* 115 (7): 1772–1780. https://doi.org/10.1016/j.rse.2011.03.008.

Ezzine, Hicham, Ahmed Bouziane, Driss Ouazar, and Moulay Driss Hasnaoui. 2017. "Downscaling of TRMM3B43 Product Through Spatial and Statistical Analysis Based on Normalized Difference Water Index, Elevation, and Distance from Sea." *IEEE Geoscience and Remote Sensing Letters* 14 (9): 1449–1453. https://doi.org/10.1109/LGRS.2017.2705430.

Fang, Bin, and Venkat Lakshmi. 2014. "Soil Moisture at Watershed Scale: Remote Sensing Techniques." *Journal of Hydrology* 516 (August): 258–272. https://doi.org/10.1016/j.jhydrol.2013.12.008.

Fang, Bin, Venkat Lakshmi, Rajat Bindlish, Thomas J. Jackson, Michael Cosh, and Jeffrey Basara. 2013. "Passive Microwave Soil Moisture Downscaling Using Vegetation Index and Skin Surface Temperature." *Vadose Zone Journal* 12 (3). https://doi.org/10.2136/vzj2013.05.0089.

Gao, Feng, William P. Kustas, and Martha C. Anderson. 2012. "A Data Mining Approach for Sharpening Thermal Satellite Imagery Over Land." *Remote Sensing* 4 (11): 3287–3319. https://doi.org/10.3390/rs4113287.

Gaur, Nandita, and Binayak P. Mohanty. 2019. "A Nomograph to Incorporate Geophysical Heterogeneity in Soil Moisture Downscaling." *Water Resources Research* 55 (1): 34–54. https://doi.org/10.1029/2018WR023513.

Hoang, L. P., H. Lauri, M. Kummu, J. Koponen, M. T. H. van Vliet, I. Supit, R. Leemans, P. Kabat, and F. Ludwig. 2016. "Mekong River Flow and Hydrological Extremes under Climate Change." *Hydrology and Earth System Sciences* 20: 3027–3041. https://doi.org/10.5194/hess-20-3027-2016.

Hornáček, Michael, Wolfgang Wagner, Daniel Sabel, Hong Linh Truong, Paul Snoeij, Thomas Hahmann, Erhard Diedrich, and Marcela Doubková. 2012. "Potential for High Resolution Systematic Global Surface Soil Moisture Retrieval via Change Detection Using Sentinel-1." *IEEE Journal of Selected Topics in Applied Earth Observations and Remote Sensing* 5 (4): 1303–1311. https://doi.org/10.1109/JSTARS.2012.2190136.

Hutengs, Christopher, and Michael Vohland. 2016. "Downscaling Land Surface Temperatures at Regional Scales With Random Forest Regression." *Remote Sensing of Environment* 178: 127–141. https://doi.org/10.1016/j.rse.2016.03.006.

Immerzeel, W. W., M. M. Rutten, and P. Droogers. 2009. "Spatial Downscaling of TRMM Precipitation Using Vegetative Response on the Iberian Peninsula." *Remote Sensing of Environment* 113 (2): 362–370. https://doi.org/10.1016/j.rse.2008.10.004.

Inamdar, Anand K., and Andrew French. 2009. "Disaggregation of GOES Land Surface Temperatures Using Surface Emissivity." *Geophysical Research Letters* 36 (2): 1–5. https://doi.org/10.1029/2008GL036544.

Jagdhuber, Thomas, Martin Baur, Ruzbeh Akbar, Narendra N. Das, Moritz Link, Lian He, and Dara Entekhabi. 2019. "Estimation of Active-Passive Microwave Covariation Using SMAP and Sentinel-1 Data." *Remote Sensing of Environment* 225: 458–468. https://doi.org/10.1016/j.rse.2019.03.021.

Jeganathan, C., N. A. S. Hamm, S. Mukherjee, P. M. Atkinson, P. L. N. Raju, and V. K. Dadhwal. 2011. "Evaluating a Thermal Image

Sharpening Model Over a Mixed Agricultural Landscape in India." *International Journal of Applied Earth Observation and Geoinformation* 13 (2): 178–191. https://doi.org/10.1016/j.jag.2010.11.001.

Jia, Shaofeng, Wenbin Zhu, Aifeng Lu, and Tingting Yan. 2011. "A Statistical Spatial Downscaling Algorithm of TRMM Precipitation Based on NDVI and DEM in the Qaidam Basin of China." *Remote Sensing of Environment* 115 (12): 3069–3079. https://doi.org/10.1016/j.rse.2011.06.009.

Jin, Yan, Yong Ge, Jianghao Wang, Yuehong Chen, Gerard B. M. Heuvelink, and Peter M. Atkinson. 2017. "Downscaling AMSR-2 Soil Moisture Data With Geographically Weighted Area-to-Area Regression Kriging." *IEEE Transactions on Geoscience and Remote Sensing* 56 (4): 2362–2376. https://doi.org/10.1109/TGRS.2017.2778420.

Kerr, Yann H., Philippe Waldteufel, Jean Pierre Wigneron, Jean Michel Martinuzzi, Jordi Font, and Michael Berger. 2001. "Soil Moisture Retrieval From Space: The Soil Moisture and Ocean Salinity (SMOS) Mission." *IEEE Transactions on Geoscience and Remote Sensing* 39 (8): 1729–1735. https://doi.org/10.1109/36.942551.

Kwon, Moonhyuk, Hyun Han Kwon, and Dawei Han. 2017. "A Spatial Downscaling of Soil Moisture from Rainfall, Temperature, and AMSR2 Using a Gaussian-Mixture Nonstationary Hidden Markov Model." *Journal of Hydrology* 564: 1194–1207. https://doi.org/10.1016/j.jhydrol.2017.12.015.

Li, Junhua, Shusen Wang, Grant Gunn, Pamela Joosse, and Hazen A. J. Russell. 2018. "A Model for Downscaling SMOS Soil Moisture Using Sentinel-1 SAR Data." *International Journal of Applied Earth Observation and Geoinformation* 72: 109–121. https://doi.org/10.1016/j.jag.2018.07.012.

Li, Shuangcheng, Zhiqiang Zhao, Xie Miaomiao, and Yanglin Wang. 2010. "Investigating Spatial Non-Stationary and Scale-Dependent Relationships Between Urban Surface Temperature and Environmental Factors Using Geographically Weighted Regression." *Environmental Modelling and Software* 25 (12): 1789–1800. https://doi.org/10.1016/j.envsoft.2010.06.011.

Merlin, Olivier, Ahmad Al Bitar, Jeffrey P. Walker, and Yann Kerr. 2010. "An Improved Algorithm for Disaggregating Microwave-Derived Soil Moisture Based on Red, Near-Infrared and Thermal-Infrared Data." *Remote Sensing of Environment* 114 (10): 2305–2316. https://doi.org/10.1016/j.rse.2010.05.007.

Merlin, Olivier, Abdelghani Chehbouni, Jeffrey P. Walker, Rocco Panciera, and Yann H. Kerr. 2008. "A Simple Method to Disaggregate

Passive Microwave-Based Soil Moisture." *IEEE Transactions on Geoscience and Remote Sensing* 46 (3): 786–796. https://doi.org/10.1109/TGRS.2007.914807.

Mishra, Vikalp, W. Lee Ellenburg, Robert E. Griffin, John R. Mecikalski, James F. Cruise, Christopher R. Hain, and Martha C. Anderson. 2018. "An Initial Assessment of a SMAP Soil Moisture Disaggregation Scheme Using TIR Surface Evaporation Data Over the Continental United States." *International Journal of Applied Earth Observation and Geoinformation* 68: 92–104. https://doi.org/10.1016/j.jag.2018.02.005.

Mohammed, I. N., J. D. Bolten, R. Srinivasan, and V. Lakshmi. 2018a. "Improved Hydrological Decision Support System for the Lower Mekong River Basin Using Satellite-Based Earth Observations." *Remote Sensing* 10: 885. https://doi.org/10.3390/rs10060885.

Mohammed, I. N., J. D. Bolten, R. Srinivasan, and V. Lakshmi. 2018b. "Satellite Observations and Modeling to Understand the Lower Mekong River Basin Streamflow Variability." *Journal of Hydrology* 564 (September). Elsevier: 559–573. https://doi.org/10.1016/J.JHYDROL.2018.07.030.

Mohammed, I. N., J. D. Bolten, R. Srinivasan, C. Meechaiya, J. P. Spruce, and V. Lakshmi. 2018. "Ground and Satellite Based Observation Datasets for the Lower Mekong River Basin." *Data in Brief* 21: 2020–2027. https://doi.org/10.1016/j.dib.2018.11.038.

Narayan, Ujjwal, and Venkat Lakshmi. 2008. "Characterizing Subpixel Variability of Low Resolution Radiometer Derived Soil Moisture Using High Resolution Radar Data." *Water Resources Research* 44 (6). https://doi.org/10.1029/2006WR005817.

Narayan, Ujjwal, Venkataraman Lakshmi, and Thomas J. Jackson. 2006. "High-Resolution Change Estimation of Soil Moisture Using L-Band Radiometer and Radar Observations Made During the SMEX02 Experiments." *IEEE Transactions on Geoscience and Remote Sensing* 44 (6): 1545–1554. https://doi.org/10.1109/TGRS.2006.871199.

Narayan, Ujjwal, Venkataraman Lakshmi, and Eni G. Njoku. 2004. "Retrieval of Soil Moisture from Passive and Active L/S Band Sensor (PALS) Observations During the Soil Moisture Experiment in 2002 (SMEX02)." *Remote Sensing of Environment* 92 (4): 483–496. https://doi.org/10.1016/j.rse.2004.05.018.Oddo, Perry, Aakash Ahamed, and John Bolten. 2018. "Socioeconomic Impact Evaluation for Near Real-Time Flood Detection in the Lower Mekong River Basin." *Hydrology* 5 (2): 23. https://doi.org/10.3390/hydrology5020023.

Peng, J., J. Niesel, and A. Loew. 2015a. "Evaluation of Soil Moisture Downscaling Using a Simple Thermal-Based Proxy – The REMEDHUS Network (Spain) Example." *Hydrology and Earth System Sciences* 19 (12): 4765–4782. https://doi.org/10.5194/hess-19-4765-2015.

Peng, Jian, Alexander Loew, Olivier Merlin, and Niko E. C. Verhoest. 2017. "A Review of Spatial Downscaling of Satellite Remotely Sensed Soil Moisture." *Reviews of Geophysics* 55 (2): 341–366. https://doi.org/10.1002/2016RG000543.

Peng, Jian, Alexander Loew, Shiqiang Zhang, Jie Wang, and Jonathan Niesel. 2015b. "Spatial Downscaling of Satellite Soil Moisture Data Using a Vegetation Temperature Condition Index." *IEEE Transactions on Geoscience and Remote Sensing*, no. Esa Cci: 1–9. https://doi.org/10.1109/TGRS.2015.2462074.

Piles, María, Adriano Camps, Mercè Vall-Llossera, Ignasi Corbella, Rocco Panciera, Christoph Rudiger, Yann H. Kerr, and Jeffrey Walker. 2011. "Downscaling SMOS-Derived Soil Moisture Using MODIS Visible/Infrared Data." *IEEE Transactions on Geoscience and Remote Sensing* 49 (9): 3156–3166. https://doi.org/10.1109/TGRS.2011.2120615.

Ulloa, Jacinto, Daniela Ballari, Lenin Campozano, and Esteban Samaniego. 2017. "Two-Step Downscaling of TRMM 3b43 V7 Precipitation in Contrasting Climatic Regions with Sparse Monitoring: The Case of Ecuador in Tropical South America." *Remote Sensing* 9 (7): 758. https://doi.org/10.3390/rs9070758.

Wang, W., H. Lu, L. R Leung, H.-Y Li, J. Zhao, F. Tian, K. Yang, and K. Sothea. 2017. "Dam Construction in Lancang-Mekong River Basin Could Mitigate Future Flood Risk From Warming-induced Intensified Rainfall." *Geophysical Research Letters* 44: 10378–10386. https://doi.org/10.1002/2017GL075037.

Yang, Guijun, Ruiliang Pu, Wenjiang Huang, Jihua Wang, and Chunjiang Zhao. 2010. "A Novel Method to Estimate Subpixel Temperature by Fusing Solar-Reflective and Thermal-Infrared Remote-Sensing Data With an Artificial Neural Network." *IEEE Transactions on Geoscience and Remote Sensing* 48 (4): 2170–2178. https://doi.org/10.1109/TGRS.2009.2033180.

Yang, Guijun, Ruiliang Pu, Chunjiang Zhao, Wenjiang Huang, and Jihua Wang. 2011. "Estimation of Subpixel Land Surface Temperature Using an Endmember Index Based Technique: A Case Examination on ASTER and MODIS Temperature Products over a Heterogeneous Area." *Remote Sensing of Environment* 115 (5): 1202–1219. https://doi.org/10.1016/j.rse.2011.01.004.

Zhan, Chesheng, Jian Han, Shi Hu, Liangmeizi Liu, and Yuxuan Dong. 2018. "Spatial Downscaling of GPM Annual and Monthly Precipitation Using Regression-Based Algorithms in a Mountainous Area." *Advances in Meteorology*. https://doi.org/10.1155/2018/1506017.

Zhao, Wei, Nilda Sánchez, Hui Lu, and Ainong Li. 2018. "A Spatial Downscaling Approach for the SMAP Passive Surface Soil Moisture Product Using Random Forest Regression." *Journal of Hydrology* 563 (June): 1009–1024. https://doi.org/10.1016/j.jhydrol.2018.06.081.

Conclusions

Alexandra Gemitzi, Nikolaos Koutsias, and Venkat Lakshmi

I T IS BEYOND ANY DOUBT that over the last 30 years satellite remote sensing offered a great advantage to the scientific community to observe and monitor the global environment. Much more is yet to come. Advances in information and space technologies are expected to promote even further the science of remote sensing in the near future and provide scientists with global data sets with unprecedent accuracy at high spatial and temporal resolution.

Nevertheless, there are certain issues that need constant attention when the reliability of our assessments is of paramount importance. Climate change has already brought important modifications to many environmental systems, both natural and human. Scientists are very fortunate to have such a precious source of information, to highlight impacted areas and determine the severity of impacts. However, not everyone can conduct accurate assessments using remotely sensed data. In our focus book, it is demonstrated that each product needs careful handling, and issues such as data filtering, computational requirements, and applicability limitations should always be carefully examined.

High resolution remotely sensed products are robust, offering decent environmental assessments, but constraints in computational requirements for the processing of such data sets may require highly efficient parallel processing algorithms, and still it is difficult to provide results for extended areas. Despite those limitations, high resolution remotely sensed products are the best suited for local and regional environmental research.

Moderate resolution remotely sensed products constitute a reasonable compromise suitable to cover extended areas at an acceptable resolution, therefore providing environmental assessments easily at the regional and continental scales.

Remotely sensed low resolution environmental data are convenient when global scale analysis is the case. For certain Earth properties, e.g. gravity field, global data sets at low resolution are still the only available source of information. Their major limitation is the coarse nature of information and consequently their limited applicability at the local scales. Within this book it is demonstrated how this type of environmental observations can be efficiently downscaled to become suitable in more detailed regional scale research.

A crucial concern of all environmental scientists who use remotely sensed information is that those data sets cannot and should not completely substitute for Earth-based environmental monitoring. In situ monitoring will always be necessary to calibrate and validate remotely sensed products. Our efforts should therefore focus, along with the promotion of remotely sensed environmental research, on the proper maintenance and expansion of in situ environmental monitoring networks.

Index

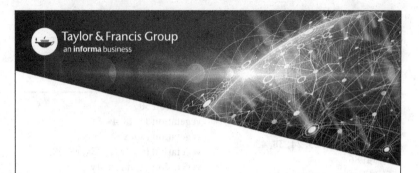

Printed in the United States
by Baker & Taylor Publisher Services